A

TEXT BOOK

ON

SURVEYING,

PROJECTIONS, AND PORTABLE INSTRUMENTS.

FOR THE USE OF

CADET MIDSHIPMEN

AT THE

U. S. NAVAL ACADEMY,

ANNAPOLIS, Md.

NEW YORK:
D. VAN NOSTRAND, PUBLISHER.
23 MURRAY ST. AND 27 WARREN ST.
1876.

CONTENTS.

PAGE.

PREFACE.

The following pages, save chapter IX, have been prepared by the officers of the Department of Astronomy, Navigation, and Surveying, of the U. S. Naval Academy. The want of a text book on the subjects discussed has long been felt in this Department; and the work is intended for the instruction of the cadet midshipmen, and not for the use of the service at large.

Lieut. R. P. Rodgers prepared Chaps. I to V, inclusive; Lt. Commander Harrington Chaps. VI to VIII, inclusive; Commodore Danl. Ammen has kindly contributed Chap. IX; Comdr. Howell prepared Chaps. X and XI; Lieut. C. P. Perkins prepared Chaps. XII and XIII.

The following is a list of the books referred to :—

Chauvenet's Trigonometry.
Chauvenet's Astronomy.
Coffin's Navigation and Nautical Astronomy.
Jeffers' Surveying.
Gillespie's Surveying.
Gillespie's Higher Surveying.
Laughton's Nautical Surveying.
U. S. Coast Survey Reports.
Mayne's Notes on Marine Surveying.
Belcher's Nautical Surveying.
General Instructions of the U. S. Coast Survey.
Mitchell's Tides and Tidal Phenomena.
Hilgard's Lecture on Tides.
Loomis' Treatise on Astronomy.
Tables and Formulæ, U. S. Corps of Engineers (Lee).
Traité de Géodésie—Puissant.
Traité des Projections—Germain.

Mr. E. R. Knorr, of the Hydrographic Office, has kindly given much assistance.

JOHN A. HOWELL,
Commander U. S. Navy,
Head of Department

U.S. NAVAL ACADEMY.
July 1, 1876.

1 kilometre = 0.62137 statute miles.

1 metre = 39.37 inches.

1 millimetre = 0.0394 inches.

Length of Gunter's chain = 66 feet = 20 links of 3.30 feet = 100 links of 7.92 inches.

Log. of mean radius of earth *in yards* = 6.8427917.

Arc of same length as radius = $57^{\circ}.2957795 = 3437'.7467 = 206264''.806$.

			In logarithms.
Arc of 180°	=	3.14159265 R.	= 0.4971498 + log. R.
" " 1°	=	0.017453293 R.	= 8.2418774 + log. R.
" " 1′	=	0.0002908882 R.	= 6.4637261 + log. R.
" " 1″	=	0.000004848137 R.	= 4.6855749 + log. R.

1 acre = 43560 square feet.

INSTRUMENTS.

CHAPTER I.

THE INSTRUMENTS.

DEFINITIONS—THE INSTRUMENTS USED—THE THEODOLITE—ITS ADJUSTMENT.

1. Topographical surveying is that branch of surveying which pertains to the process of ascertaining and representing upon a plane surface the contour or figure of any portion of the surface of the earth. (Webster.)

Plane surveying is surveying as carried on under the supposition that the surface of the earth is a plane in distinction from geodesic surveying. (Webster.)

Geodesy is that branch of applied mathematics which determines by means of observations and measurements, the figure and areas of large portions of the earth's surface, or the general figure and dimensions of the earth; or that branch of surveying in which the curvature of the earth is taken into account, as in the survey of states or long lines of coast. (Webster).

2. The instruments for measuring angles which are generally used in the survey of a harbor are, the Theodolite, Sextant, and Plane Table. All officers are sufficiently acquainted with the handling of a sextant. To use the plane table to advantage requires much more practice and experience, in addition to a theoretical knowledge, than to get the same result by the Theodolite or Sextant, and consequently a treatise upon it will not to be attempted in this work. Those desirous of obtaining information on the subject are referred to the Coast Survey

Report of 1865, or to a pamphlet issued from Coast Survey office, which contains the same matter.

The Theodolite is the instrument chiefly used, and a brief description of this instrument together with its adjustments is here given.

The instruments for measuring distances are the chain or steel tape, wooden rods or carefully measured lengths of line—and the telemeter.

In conducting the hydrographic portion of the work, to the above are added the lead, tide-guage, and instruments for measuring the velocity of the current.

THE THEODOLITE.

3. The Theodolite appears in a variety of forms; its purpose is to measure horizontal, and sometimes, vertical angles.

It consists essentially of a telescope which has a motion about an axis (horizontal) which rests in Ys in two pillars, F F, which are perpendicular to the axis of rotation of the telescope (See Plate 1.) These pillars are fixed at right angles to a plate, K, which turns upon a vertical axis and to which is attached a vernier. Around that just mentioned is a second plate, K_1. The latter is graduated, and is concentric with the first.

The instrument is mounted upon a tripod T. Three screws, P, with milled heads, which bear upon the plates are used in connection with spirit levels, N, upon them, for making them level. A plumb bob hangs directly under the centre of the instrument.

Sometimes a small graduated circle, G, and level with vernier, H, is fixed upon, and at right angles to, the axis of rotation, E, for the measuring of vertical angles. In the common focus of the object and the eye glasses is placed the diaphragm, which contains the cross-wires; usually one vertical and the other horizontal, bisecting each other in the optical axis.

4. *Definitions* (See Chauvenet's Astronomy).

The *Axis of Rotation* is an imaginary line passing through the centre of the pivots of the axis.

The *Axis of Collimation* is an imaginary line drawn from the optical centre of the object glass perpendicular to axis of rotation.

The *Line of Sight* is an imaginary line drawn from the optical centre of object glass to the (in case there is but one) vertical thread.

The angle which the line of sight makes with the axis of collimation is called the *Collimation error.*

THE ADJUSTMENTS OF THE THEODOLITE.

5. In the measuring of horizontal angles by the Theodolite it is required,

1. That the Azimuth plates shall be horizontal in all positions when turned about their axis;

2. That the *Line of Sight* shall coincide with the *Axis of Collimation;*

3. That the Line of Sight shall move in a vertical plane, when the telescope is turned about its axis. In order that these conditions may be fulfilled the following adjustments are necessary :

6. 1st. To cause the Azimuth plates to be horizontal in all positions.

Turn the vernier plate which carries the levels till one of them is parallel to *one pair* of levelling screws. Bring the bubble of each to the middle of its tube by means of the three screws. Then turn the plate around slowly 180°, and if the bubbles remain in the middle of the tubes, then the plates are horizontal. If the bubbles have moved from the centre, then the levels themselves must also be adjusted. The error to be adjusted is partially occasioned by the level not being perpendicular to the so called vertical axis. Let AB represent the plane of one of the levels, and CD the central line of the vertical axis, which is shown as making an acute angle with the plane AB on the right side. Fig. 1 represents the bubble of level as in the centre of tube, Fig. 2 represents the plates turned 180°, the central line of axis is supposed to remain unmoved, and the acute angle will now appear on the left side of it, and the bubble will rise to the higher end of the tube. The error of the level is half the motion of the bubble; for half the motion was caused by the plane of the level not being perpendicular to the vertical axis, and half by the axis itself not being vertical. Therefore raise one end of the level tube, by means of screws which fasten it to the plate, till the bubble moves half way back to the centre of the tube, and then bring it entirely back to the centre by means of the levelling screws under the azimuth plates. Now turn the plates again through 180°, and

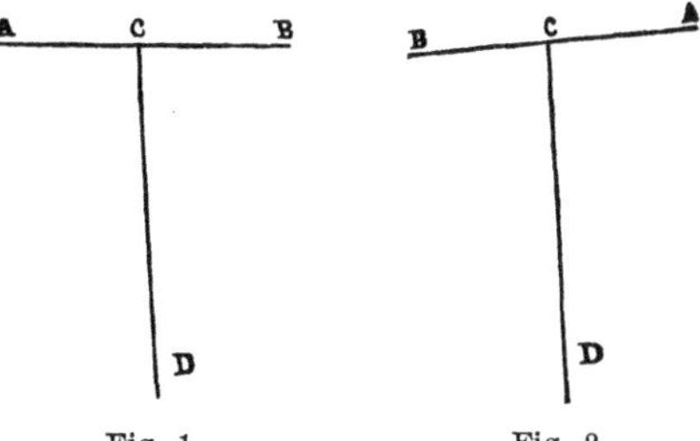

Fig. 1. Fig. 2.

should the bubble now remain in the centre of the tube, then the level is adjusted and the plates horizontal; because the plane of the plates is at right angles to their axis. If the bubble does not remain in the centre, this operation must be repeated till it does. The second level may then be adjusted directly by *its* adjusting screws. It is also necessary to see whether the bubbles remain in the centre of tubes when the graduated circle is moved around its axis. If they do not, the axis of the two plates are not coincident with each other, which can only be remedied by the instrument maker.

7. 2d. The adjustment for instrumental parallax. Instrumental parallax is an apparent movement of the cross-hairs about the object to which the line of sight is directed, occasioned by a movement of the eye from right to left or the contrary. It is caused by the image of the object and the cross-hairs not being precisely in the common focus of the eye piece and the object glass. To eliminate this error, move the eye piece out or in till the cross-wires are distinctly seen, and then move the whole system, by means of the ratchet and pinion (C, Fig. 1, Plate I.) which moves the eye tube, V, until the image of the object is most distinct. The cross-hairs and image will then be in the common focus.

8. 3d. To cause the so-called vertical wire of the diaphragm to be at right angles to the axis of rotation.

This may be done at the same time that the collimation error is being corrected. Point the telescope to some well-defined object, and bisect it with the vertical wire. Give the telescope a slight motion about its axis of rotation, and if the vertical wire continues to bisect the object it is at right angles to axis of rotation. If it does not, loosen the screws, D, which secure the diaphragm and turn it till the wire bisects the object throughout its entire length. If the axis of rotation is horizontal, this wire will then be vertical.

9. 4th. To cause the line of sight and the axis of collimation to coincide.

If the axis of rotation of the telescope can be taken out of its supports and turned end for end, then the collimation error can be determined in the following manner. Set the instrument up, and having made each of the before mentioned adjustments, place the zeros of azimuth plates in coincidence, and direct the line of sight to some well defined and distant object. Then take the telescope out of the Ys, and reverse the axis. If the line of sight falls precisely upon the same object, then there is no collimation error. If it should not fall upon the object after reversal, turn the line of sight upon it by means of the tangent screw of the upper plate, L (keeping the lower plate fixed), and note the reading. The difference of the readings of the two directions of the telescope will be double the collimation error, as will be seen from the annexed figures. Let y, y' represent the axis of the telescope in first position, and let the point of sight fall upon the point s, and the axis of collimation upon c. Let the axis of the telescope be reversed and take position y_1' y_1. The line of sight will fall upon s', as much

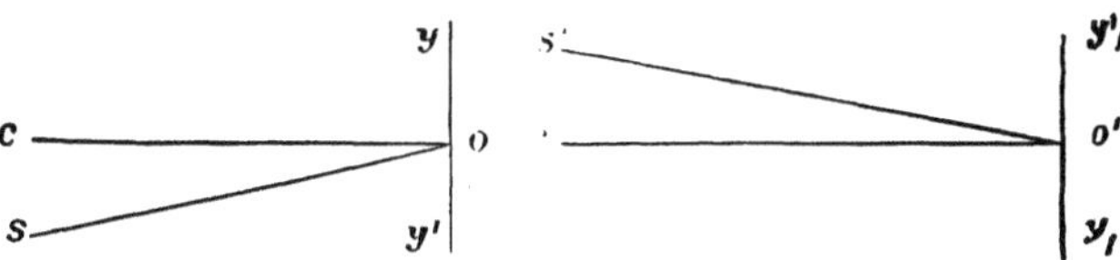

to the right of c as it was before to the left. Consequently the collimation error is equal to one half the difference of direction of the lines os and o' s', which is one half the difference of the readings already noted. If after reversal, the vertical wire in the diaphragm be moved by the screws D till the line of sight falls midway between its first and second positions, then will the line of sight and the axis of collimation coincide. Probably the distance through which the diaphragm is to be moved will not be accurately estimated, so that the operation must be repeated till the collimation error is reduced to an inappreciable quantity.

The axis of rotation of the telescope in many theodolites is not capable of being reversed; we must then proceed in the following manner to eliminate the collimation error. See following figure.

Set the theodolite up in the middle of a level piece of ground, and make the foregoing adjustments. Place a small but well-defined mark (at B) at a distance as large as possible from the instrument (at A). Turn the telescope toward it and perfect the intersection of the cross-wires and mark (B) by means of the tangent screw, clamp the plates firmly.

Transit the telescope (that is turn it about its axis of rotation till it points in the opposite direction, the axis of rotation remaining fixed) and place another mark at the same distance from the instrument as (C), and in the line of sight. If the line of sight and axis of collimation do not coincide, this second mark will be placed at C, and not at E, in the prolongation of A B. To ascertain whether B and C are in line, unclamp the instrument, and without touching the telescope, turn the vernier plate around till B comes into its field. Perfect the intersection of the cross-wires and B, and clamp the plate firmly. Then transit the telescope, and the line of sight will (if C A X is the collimation error) fall upon some point D, as much to the right of E as C was to the left of it.

The figure will show the position of the line of sight in reference to the axis of collimation, when the telescope is in the four positions mentioned. Let y y'' represent the axis of rotation: XX the axis of collimation of telescope in 1st and 2d positions, y_1' y_1'' X′X′ the

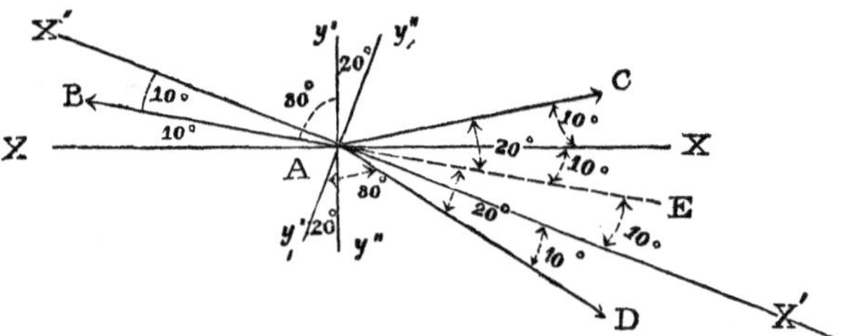

same in 3d & 4th positions. In this case 10° has been taken as the collimation error, and it will be seen that when the line of sight (represented by arrow-headed lines) has been directed to B, and the telescope has been transited, it will fall upon C, 20° to the left of AE, the prolongation of AB. Now when the vernier plate is turned so that the line of sight falls upon B, the axis of rotation will assume the position y_1' y_1'' and the axis of collimation that of X′ X′. Now when the telescope is again transited the line of sight falls upon D 20° to the right of E. By looking at the figure it will be seen that the line of sight makes an angle of 80° and 100° with the axis, that of 80° being made on side of axis marked y' or y_1', that of 100° on side marked y'' or y_1''; and that although the axis cannot be taken out of the Ys and reversed, yet in reality the telescope has been directed to the same object B with the axis pointing in opposite directions, and consequently the collimation error is shown just as exactly as when the axis was reversible, excepting that in this case the angle C A D is four times the error. To eliminate the collimation error, move the telescope through three quarters of the arc, or D X, and then move the diaphragm till the vertical wire falls upon C. The collimation error will, if the telescope and diaphragm have been properly moved, be zero. The arc C D will be obtained by taking the difference of readings of the circle when the line of sight falls upon C and D respectively, the circle having remained fixed.

10. 5. To cause the line of sight to move in a vertical plane.

Suspend a long plumb line from some high point. Set the instrument up near this line, and *level it carefully.* Direct the telescope to the plumb line, and if the intersection of the cross-hairs remains upon the line throughout its entire length, then the line of sight moves in a vertical plane. Or, the instrument having been levelled, place a basin of mercury between the instrument and some fixed object of considerable altitude, such as the top of a flag-staff or tower; turn the telescope down to bring the reflection upon the intersection of the cross-wires, then move the telescope, and if, when directed toward the object itself, the cross-wires coincide directly with the object, the line of sight has moved in a vertical plane.

If, however, the above conditions are not fulfilled, then the axis of rotation is not horizontal; or in other words, the pillars which support the axis are not of equal length, measuring from the plane of the plates. For the line of sight has been made to coincide with the axis of collimation, and were the axis of rotation horizontal it would necessarily describe a vertical circle. With those theodolites on which there is any arrangement for moving the supports of the axis of rotation (see Fig. 1, Plate I.), the adjustment may be made by moving the levelling screws in connection with the striding level, which is placed upon the axis. But in the ordinary form of theodolite there is no such arrangement; so that if the error of level of the axis of rotation (while the plane of the plates is horizontal) is great, the instrument should be sent to an instrument maker; but if the error is but slight the bearing of the support of the higher end of the axis may be carefully filed down. This error will rarely be but small in any but an old instrument.

The theodolite furnished by the Bureau of Navigation (Plate II.) differs from the one already described in the following respects: It has but one moveable azimuth plate (the upper). It has no vertical circle. It is levelled by means of four levelling screws acting on a ball in a socket joint.

As it has but one moveable azimuth plate, angles cannot be repeated, nor can a line be run by the back angle. It is a very convenient instrument for the purposes of an ordinary survey, and is furnished with a telescope of considerable power.

CHAPTER II.

BASES.

SELECTION OF BASES—DIRECT MEASUREMENT OF—REDUCTION OF INCLINED BASE TO HORIZONTAL MEASURE—BROKEN BASES—MEASUREMENT OF, BY SOUND—BY MAST-HEAD ANGLES—DETERMINATION OF AZIMUTH OF BASE—FIXING THE LATITUDE AND LONGITUDE OF A BASE STATION.

MEASUREMENT OF THE BASE-LINE.

11. Before any distance can be measured trigonometrically the length of one straight line of the triangulation must be mechanically measured. And as this measurement is more or less exact, so will all other measurements deduced from it be more or less accurately determined.

If an error in the length of the base be represented by $d\,b$; then $d\,a = d\,b \frac{\sin A}{\sin B}$; $d\,c = d\,b \frac{\sin C}{\sin B}$($da$ and dc representing the consequent errors in the other two sides). If $B = 30°$, since $\sin 30° = \frac{1}{2}$; $d\,a$ or $d\,c$ *may* equal $2\,d\,b$. Therefore the error of the sides may be double that of the base, even where the triangles are not ill-conditioned.

Measuring the base line for an extended and very accurate survey is effected by metallic rods, accurately graduated, which are carefully levelled, and the several distances are determined by means of micrometers; but for the purpose of an ordinary survey the base-line may be sufficiently well measured by means of a 100 feet steel tape in the following manner:

12. Before commencing the measurement the tape should be carefully compared with a standard measure, and the error noted, and the ground over which the line is to pass should be made even. Too much care and attention cannot be paid to this.

Set a theodolite up over one extremity of the base, so that the plumb is directly over the centre of the station, and after having levelled the instrument, point the telescope toward the other extremity of the base line, at which should be placed a signal.

Have a number of poles, or boarding pikes, with flags secured to one end, which are to be placed vertically in the line, at intervals of about 100 or 200 feet, by the observer at the theodolite, who, by motion of his hand to the right or left, or by some other preconcerted signal, causes the flagmen to move until the flag poles are bisected by the cross-wires of the instrument.

The measuring party consists of two men who carry the tape and of two others who carry small pieces of plank, or benches, upon which the ends of the tape are marked. A good form of bench is a plank about one foot square, with four short, pointed legs, which may be firmly planted in the ground by the pressure of the foot after the measure has been approximately adjusted both for distance and direction. One end of the tape is held directly over the centre of the station from which the measurement begins, and the other, being carried forwards, is brought into the line of the flags by the man holding the rear end of the tape. A spring balance is hooked into the forward end of the tape and a pressure of 8 or 10 pounds is brought upon it, drawing the tape taut along the line. One of the men with the benches places his bench flush with the ground, under the tape near its end, and making a mark opposite the 100 foot mark of the tape, which is numbered 1, he holds the bench securely in its place. The tape carriers now move ahead; the carrier to the rear places his end of the tape on the mark on the bench indicating the point which the forward end of the tape before reached; the forward carrier moves into the line of the flags, draws the tape taut and in the line, applies the same pressure, by spring balance, as before; a second man with bench places it under the tape, marking the point where the 100 foot mark falls No. 2. The tape is again carried forward, aligned, and the distance marked as before. As the measuring party passes each flag, the flag bearer will pass on beyond the last flag and will be aligned in his new position by observer at theodolite.

This process is continued until the other extremity of base line is reached, noting the number of feet and tenths of a foot between the last point marked by the tape and the centre of station to be reached. Then the length of tape ± error of tape (as determined by comparison with standard before and after measurement) multiplied by number of lengths of tape + the distance between point last reached by a whole length of the tape and the extremity of the base line equals length of base line.

13. Not being able to obtain a steel tape, take a piece of iron wire, about $\frac{1}{8}$ inch in diameter, about 60 yards (or any desired number) in length. This distance is carefully measured off by standard measures, and both extremities are marked on the heads of copper nails driven in a plank, level with ground. The wire, which has been straightened, and otherwise prepared, and also provided with a loop at either end, is then stretched between two marks by a chain staff at the rear end and a weight of 40 pounds applied to the other by a spring balance. In this condition, and after repeated trials, the measured distance is transferred to, and marked upon, the wire by a fine cut near each loop. This having been completed, proceed in the same manner as explained when the steel tape was used.

To show the accuracy of the above methods reference may be had to Coast Survey Report, 1868, in which is stated that a base line was measured, using wire $\frac{1}{8}$ inch in diameter and 60 metres in length, with following results:

Length of wire = 60·01646 metres.

By Wire measurement,	A to B =	82 wires × 60·01646 m. =	4921·3495 m.
" Base Apparatus,	A to B =		= 4921·1843.
" Wire,	B to C =		= 4261·1685.
" Base Apparatus,	B to C =		= 4261·3225.

14. To reduce the length of an inclined base to horizontal measure.

When the inclination of the base is considerable, or it is measured by chain or tape, which cannot be placed horizontally, it is necessary to reduce the base to its horizontal length.

Let B = length of base on an inclined plane.

b = " " reduced to horizontal plane.

θ = inclination of base to horizon.

Then $b = B \cos \theta$ (1)

But as θ is generally a small angle, and need not be known with extreme precision, it is better to compute the excess of B over b. Supposing θ to be given in minutes, we have from

$$(1)\ B - b = B - B\cos\theta = B(1 - \cos\theta) = 2B\sin^2 \tfrac{1}{2}\theta$$

$$= \tfrac{1}{2} B\, \theta^2 \sin^2 1'$$

$$= \frac{\sin^2 1'}{2}\, \theta^2 B = 0.00000004231\, \theta^2 B, \text{ or by logarithms.}$$

Log. (B–b) = const. log. 2·626422 + 2 log. θ + log. B, (2)
which gives correction to be subtracted from B to get b.

THE BROKEN BASE.

15. The character of the ground will sometimes not allow a base of suitable length to be measured directly, but will permit of measuring two parts, making a very obtuse angle with each other, constituting what is known as a "Broken Base." It is necessary to measure the two portions a and b, and the included angle C with great care; we can then compute the base by expression here deduced:

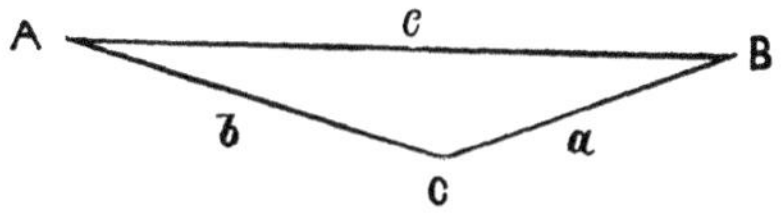

From Trig. $c^2 = a^2 + b^2 - 2\,a\,b \cos C$ (1)

Let $\theta = 180 - C$.

Then $c^2 = a^2 + b^2 + 2\,a\,b \cos \theta$ (2)

From plane Trig. (406) (developing cos x into a series)

$$\cos x = 1 - \frac{x^2}{1.2} + \frac{x^4}{1.2.3.4} - \&c.,$$

substituting θ for x, and since θ is very small,

$$\text{Cos } \theta = 1 - \frac{\theta^2}{2}$$

This will be sufficiently correct if C is not less than 170°.

$$c^2 = a^2 + b^2 + 2\,a\,b - a\,b\,\theta^2 = (a + b)^2 - a\,b\,\theta^2 = \quad (3)$$

$$(a + b)^2 \left(1 - \frac{a\,b\,\theta^2}{(a + b)^2}\right)$$

$$c = (a + b)\left(1 - \frac{a\,b\,\theta^2}{(a + b)^2}\right)^{\frac{1}{2}} \quad (4)$$

Developing the quantity under the radical to $\frac{1}{2}$ power, by the bynomial theorem, and neglecting the terms after the second, it becomes

$$1 - \frac{1}{2}\,\frac{a\,b\,\theta^2}{(a + b)^2} + \&c.$$

substituting for θ minutes, θ sin 1′ we have

$$c = a + b - \frac{a\,b\,\theta^2 \sin^2 1'}{2\,(a + b)} \quad (5)$$

$$\frac{\sin^2 1'}{2} = 0{\cdot}00000004231$$

$$c = a + b - 0{\cdot}00000004231 \times \frac{a\,b\,\theta^2}{a + b} \quad (6)$$

$$\log.\ 0{\cdot}00000004231 = \bar{2}{\cdot}6264222.$$

Example taken from C. S. R. 1868.

$$\left.\begin{aligned} a &= 4957.931162 \\ b &= 4307.587099 \end{aligned}\right\} \text{metres.} \qquad a + b = 9265.518261$$

$$\begin{aligned} C &= 177^\circ 37' 59''.82 \\ \Theta &= 2^\circ 22' 0''.18 \\ \Theta &= 142'.C03 & 2 \text{ log. } 4.3047492 \\ a &= 4957''93116 + & \text{log. } 3.6953005 \\ b &= 4307.5871 + & \text{log. } 3.6342341 \\ a + b &= 9265.5182 + & \text{ar. co. log. } 6.0331303 \\ & & \text{const. log. } 2.6264222 \\ & & \text{log. } 0.2938363 \end{aligned}$$

$$\begin{aligned} & -1.967247 \\ & 9263.551014 = \text{length of base.} \end{aligned}$$

TO FIND A PORTION, x, OF A STRAIGHT LINE WHICH CANNOT BE DIRECTLY MEASURED.

16. In the survey of a harbor it frequently happens that a sandy

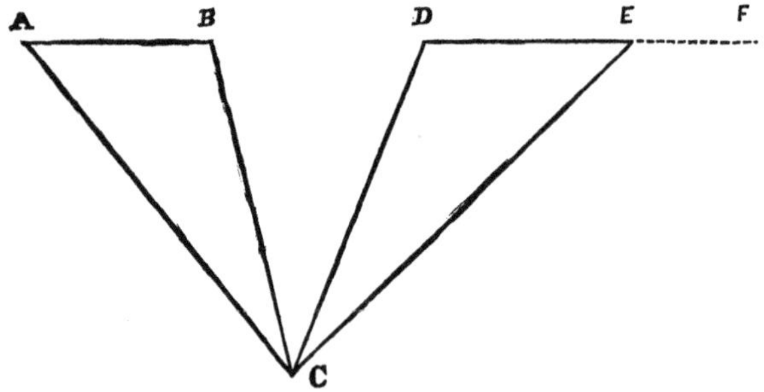

beach, or other suitable place for measuring a base, may be intersected by a stream too wide and deep, or a ravine too steep, for convenient measurement by rods or chains, &c. In this case it is required to find that portion x = BD, knowing the other two portions of the base by actual measurement. AB=a, DE=b, and also the angle ACB=α, ACD=β, ACE=γ. Then CBD=A+α; CDE= A +β CEF=A+ γ.

From triangle ABC, $\dfrac{BC}{a} = \dfrac{\sin A}{\sin \alpha}$, Δ ADC, $\dfrac{DC}{a+x} = \dfrac{\sin A}{\sin \beta}$

Dividing one by the other, we have

$$\frac{BC}{DC} = \frac{a \sin \beta}{(a+x) \sin \alpha} \tag{1}$$

$$\text{From } \Delta \text{ BEC } \frac{BC}{b+x} = \frac{\sin (A+\gamma)}{\sin (\gamma - \alpha)}$$

$$\Delta \text{ DCE, } \frac{DC}{b} = \frac{\sin (A+\gamma)}{\sin (\gamma - \beta)}$$

$$\therefore \frac{BC}{DC} = \frac{\sin(\gamma - \beta)\,(b + x)}{\sin(\gamma - \alpha)\,.\,b} \qquad (2)$$

combining (1) and (2), we have

$$\frac{a \sin\beta}{(a + x)\sin a} = \frac{(b + x)\sin(\gamma - \beta)}{b\sin(\gamma - \alpha)} \qquad (3)$$

clearing of fractions

$$a\,.\,\sin\beta\,.\,b\sin(\gamma - \alpha) = (a + x)\,(b + x)\sin\alpha\sin(\gamma - \beta) \qquad (4)$$

$$\frac{a\,.\,b\sin\beta\sin(\gamma - \alpha)}{\sin\alpha\sin(\gamma - \beta)} = (a + x)\,(b + x) = ab + (a + b)\,x + x^2 \qquad (5)$$

To solve this equation of 2d degree with reference to x, make

$$\tan^2\varphi = \frac{4\,a\,b\sin\beta\sin(\gamma - \alpha)}{(a - b)^2\sin\alpha\sin(\gamma - \beta)}$$ which substituted in (5) gives

$$\tfrac{1}{4}(a - b)^2 \times \tan^2\varphi - a\,b = x^2 + (a + b)\,x, \qquad (6)$$

Completing the square of 2d member,

$$\tfrac{1}{4}(a - b)^2 \times \tan^2\varphi - a\,b + [\tfrac{1}{2}(a + b)]^2 = x^2 + (a + b)\,x + [\tfrac{1}{2}(a + b)]^2, \text{ and} \qquad (7)$$

$$\sqrt{\tfrac{1}{4}(a - b)^2\tan^2\varphi - a\,b + \tfrac{1}{4}(a + b)^2} = x + \tfrac{1}{2}(a + b)$$

$$x = -\tfrac{1}{2}(a + b \pm \sqrt{\tfrac{1}{4}(a - b)^2\tan^2\varphi - a\,b + \tfrac{1}{4}(a + b)^2} \qquad (8)$$

$$= -\tfrac{1}{2}(a + b) \pm \sqrt{\tfrac{1}{4}(a - b)^2\tan^2\varphi + \tfrac{1}{4}(a - b)^2}$$

$$= -\tfrac{1}{2}(a + b) \pm \tfrac{1}{2}(a - b)\sqrt{\tan^2\varphi + 1}$$

but $\sqrt{\tan^2\varphi + 1} = \sec.\ \varphi = \dfrac{1}{\cos\varphi}$

$$\therefore x = -\tfrac{1}{2}(a + b) \pm \frac{a - b}{2\cos\varphi} \qquad (9)$$

Then base line $= a + x + b$

THIS EXAMPLE IS TAKEN FROM C. S. REPORT, 1868.

Let AB represent a section of the regularly measured base, about equal to the distance to be crossed, with its ends carefully marked.

Recommence the measurement on the other side of creek or bog, and mark off a similar distance ED, all in the line of the base. Set up and adjust the theodolite on firm ground at C and in sight of the stubs at A, B, E, and D. At the intersection of the cross-lines on the tacks at these four points, or in line to C, insert a short wire, or nail, or the smallest object distinctly visible from C. Then measure the angles ACB, ACD, and ACE, and for verification, the angle BCE.

Crossing of Rudie Creek.

Let AB $= a = 90^m.0242$ = distance corrected for inclination and temperature.

DE $= b = 120^m.0316$ = distance corrected for inclination and temperature.

BD $= x =$ the unmeasured distance.

ACB $= \alpha = 19° \; 41' \; 44''.56$ = mean of observations.

ACD $= \beta = 49° \; 02' \; 29''.77$ = mean of observations.

ACE $= \gamma = 75° \; 22' \; 02''.56$ = mean of observations.

BCE $= \gamma - \alpha = 55° \; 40' \; 14''.50$ = mean of observations.

$$\tan^2 \varphi = \frac{4\,ab}{(a-b)^2} \cdot \frac{\sin \beta \sin (\gamma - \alpha)}{\sin \alpha \sin (\gamma - \beta)}$$

$$x = -\frac{a+b}{2} \pm \frac{a-b}{2 \cos \varphi}$$

$\gamma - \alpha$	$= 55° \; 40' \; 16''.25$ *l.* sin.	9.9168826
β	$= 49° \; 02' \; 29''.8$ *l.* sin.	9.8780538
$\gamma - \beta$	$= 26° \; 19' \; 32''.8$ *l.* cosec.	0.3531314
α	$= 19° \; 41' \; 44''.6$ *l.* cosec.	0.4723380
log a	$= 90^m.0242$	1.9543593
log b	$= 120^m.0316$	2.0792956
log 4		0.6020600
co. log $(a-b)^2$		7.0455432
$\tan^2 \phi$		2.3016639
$\tan \phi$		1.1508319
ϕ	$= 85° \; 57' 29''.7$	
log $(a-b)$		1.4772284
ar. co. log cos		1.1519135
ar. co. log 2		9.6989700
$\frac{a-b}{2 \cos \phi}$	$212^m.8687$	2.3281119
$\frac{a+b}{2}$	$105^m.0279$	
$x =$	$107^m.8408$	

For verification, change the position of C, and from the new point observe the same angles, and from them recompute x. In the above case the different values x, are found to coincide.

Measurement of Base by Sound.

17. Sometimes circumstances will not admit of the direct measurement of the base, when it may be approximately determined by means of the velocity of sound. This method is applicable in the survey of an extensive shoal or reef, which offers no points above water for observation, or upon which to land.

The observers, at either end of the base, are provided with howitzers or other guns, thermometers, and watches. Having been previously warned, the observer at the opposite extremity of the base from which the gun is fired, notes the time elapsed from the instant of seeing the flash and that of hearing the first sound of the report. The guns are fired alternately from the two extremities at least three times. The means of these observations gives the time which it has taken the sound to pass over the distance, which may be calculated by the following formula:

$D = V \times T$ $\quad$ V = velocity of sound calculated as below. T = mean of times noted.

$$V = 1089.4\sqrt{1 + (t^\circ - 32^\circ) \times 0.00208}$$

in which t° is the height of the thermometer ; 1089.42 being the velocity of sound in feet at 32° Fahrenheit.

The velocity of sound through the air is independent of the barometric pressure, and experiments show it to be sensibly unaffected by its hygrometrical state; by the original direction of the sound, whether, for instance, the muzzle of a gun is turned one way or the other, or by the nature and position of the ground over which the sound is conveyed. It is affected by the wind ; but in ordinary cases, likely to be selected for experiment, its influence would be almost inappreciable. (Lee's Tables and Formulae, 1873, Corps of Engineers, U. S. A.)

Determination of Length of Base by Mast-Head Angles.

It frequently may occur that no suitable tract of land can be found for the accurate measurement of a base ; and frequently, when taking into consideration the degree of accuracy to be desired, that an approximate value for the length of a base may be considered all that is necessary. Such an approximation may be determined from the measurement of mast-head angles, the perpendicular height of the mast being carefully taken from the draft of the ship. As the height of the observers will be somewhat raised above the level of the water it appears from the figure that the triangle A B C is not a right triangle and it would seem that it would be necessary to find some other expression for the determination of the distance x

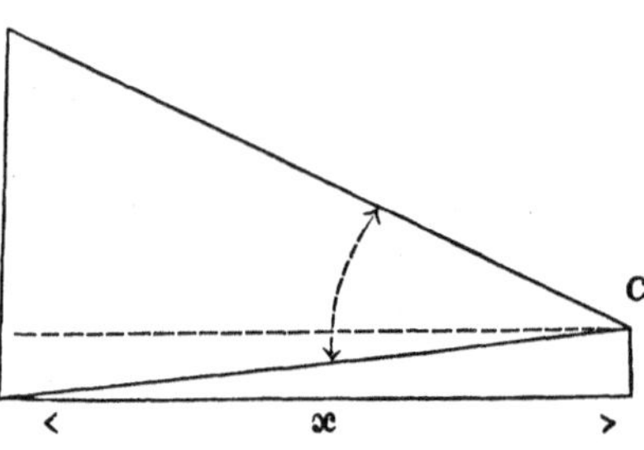

than $\tan\theta = \frac{AB}{BC}$ in which B C is taken equal to $x = AB \cot\theta$. But by investigation, as will be shown in the next Art., it will be found that the difference between BC and x will, under most circumstances, be inappreciable.

But this distance x is the distance between the ship's mast and the observer in the boat or ashore, and as the ship is liable to swing and thus change her position, it is desirable to find from the distance thus determined the distance between some two objects on the shore.

It may be done in this way. Choose two points A B in sight from each other, and situated so that the line A B will make a convenient base line, from which to carry on the triangulation. Send an observer to each station. Then at a preconcerted signal (while the ship is stationary) let the observer at A measure the mast-head angle and the angle C A B; the observer at B the angle A B C; and at C (the foot of mast angled upon) measure the angle A C B. From the mast-head angle compute the distance x, taking into consideration the height of the observer at A, if he be considerably elevated. Then from the

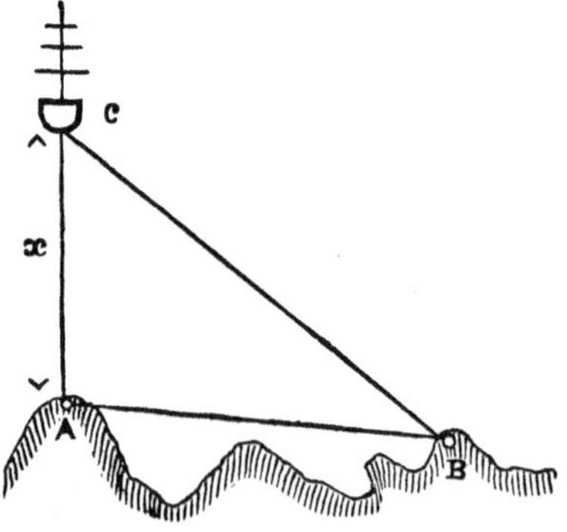

triangle A B C compute the distance A B. Of course the third observer at C is not absolutely necessary, but his angle furnishes a check upon the other two.

19. To show that a slight elevation of the eye above the water when measuring mast-head angles need not be considered in computing the distance. The angle BDA = BCA: therefore when we use formula

$$x = \text{AB cot. BCA},$$

we determine AD instead of AC = AP.

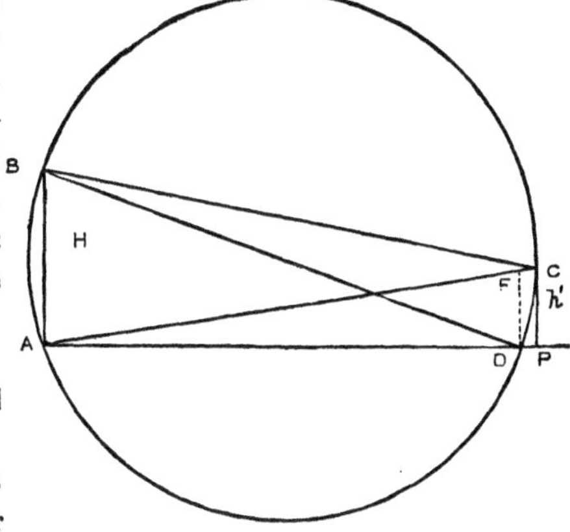

If DC is a small arc, less than 1°, we may consider it straight; then the triangle DCP is similar to ABD, since the sides are mutually perpendicular to each other.

$$\therefore \text{DP} : \text{PC} = \text{AB} : \text{AD or DP} = \frac{\text{PC} \times \text{A}}{\text{AD}} = \frac{h' \times \text{H}}{\text{AD}}$$

If the diameter of the circle ABCD is 3,000 feet, then 1° = 26.18 feet, or $h' = 26.+$ feet. So that were the eye elevated 20 ft. and the ship's mast, or H, 180 feet high, DP, or the error in the computed distance, would only be 1.2 feet. With the diameter of the circle 700 feet, the length of 1° = 6.1 feet, so that at this distance an elevation of the eye of 6 feet, with the mast 180 ft. high, would only make an error 1.5 ft. in the computed distance. From which we see that when the product of H and h' equals the distance, the error is but one unit of measure, *provided* the arc CD does not exceed 1°, which will be ensured when CAD $<\frac{1}{2}°$.

Should the height of the observer, or h', be so great that the arc CD is larger than 1°, the more rigorous formula taken from Jeffers' Surveying (page 99) may be used. Observing the same notation,

we have $\tan a = \dfrac{\text{H} - h'}{x}$; $\tan \phi = \dfrac{h'}{x}$

$$\tan \theta = \tan (a + \phi) \tag{1}$$

$$\text{Trig. (123)} \tan \theta = \tan (a + \phi) = \frac{\tan a + \tan \phi}{1 - \tan a \tan \phi} =$$

$$=\frac{\mathrm{H}x}{x^2-h'(\mathrm{H}-h')}=\frac{\mathrm{H}}{x-h'\dfrac{(\mathrm{H}-h')}{x}} \qquad (2)$$

$$\text{but } \tan\alpha=\frac{\mathrm{H}-h'}{x} \qquad \therefore \qquad \tan\theta=\frac{\mathrm{H}}{x-h'\tan\alpha}$$

$$x\tan\theta-h'\tan\alpha\tan\theta=\mathrm{H}$$

$$x=\frac{\mathrm{H}}{\tan\theta}+h'\tan\alpha \qquad (3)$$

To find $h'\tan\alpha$ take the approximate value.

$x=\dfrac{\mathrm{H}}{\tan\theta}$ substituting this in $\tan\alpha=\dfrac{\mathrm{H}-h'}{x}$

we have $h'\tan\alpha=\dfrac{h'(\mathrm{H}-h')}{\dfrac{\mathrm{H}}{\tan\theta}}$ which substituted in (3)

$$\text{gives } x=\frac{\mathrm{H}}{\tan\theta}+\frac{h'(\mathrm{H}-h')}{\dfrac{\mathrm{H}}{\tan\theta}} \qquad (4)$$

Under ordinary circumstances, as has been before shown, the second term will be too small to be considered.

The Azimuth or Astronomical Bearing of a Side of the Triangulation.

20. In order that the different points of the triangulation shall, when projected upon the chart, have the same relative positions upon it as they have upon the earth's surface, it is necessary to determine the true or astronomical bearing of one of the sides of the triangulation. It is usually most convenient to determine that of the base line.

When a theodolite is available the following method may be used. Place the theodolite at one extremity of the base, directly over the centre of the station and make the necessary adjustments. Place the zeros in coincidence; unclamp the lower plate and bisect the signal at other extremity of base by the cross-wires of telescope. Clamp lower plate firmly. Unclamp upper plate and turn the telescope so as to bring the sun (or other heavenly body) into its field. Have an assistant at hand to mark the time by chronometer, whose error is accurately known (or watch which has already been compared with chronometer.) Clamp the upper plate and, by means of its tangent screw and the tangent screw of the vertical circle, cause the vertical wire in telescope to become tangent to the sun's disc. Note the time, and read the angle. Repeat this operation several times in quick succession. Then the mean of the angles will correspond to the mean of the times, provided the time elapsed between the first and last time is not too great for the supposition that the change of azimuth is proportional to the change of time. By means of formulae given hereafter compute the azimuth of the sun. Then the angle measured by theodolite (to which must be added or subtracted a correction for the sun's semi-diameter) plus or minus the true azimuth of the sun will be the azimuth or astronomical bearing of the base line. If the sun is to left of base and is east of meridian the angle between the sun and base must be added to the sun's azimuth to get true bearing of base. If sun is to the west of meridian in the above case, the angle must be subtracted. All this refers to N. Latitude. The observation should be made when the heavenly body is near the prime vertical, provided it has not too great an altitude at that time.

In this method the local time must be accurately known by chronometers, or if these are regulated to Greenwich time, the Longitude must have been accurately determined.

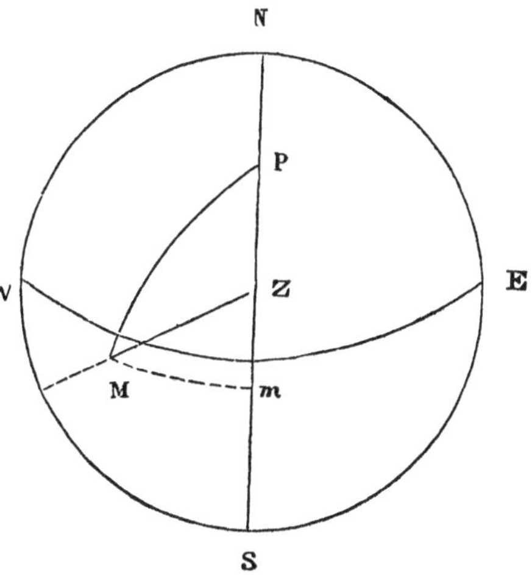

21. In fig. L = Lat, usually regarded as positive. d = declination of body; positive or negative, as it is of same or opposite name to Latitude. t = Hour angle of the body. φ' = an auxiliary, the declination of the foot of the perpendicular let fall from the body upon the meridian. h = body's true alt.

From trigonometry $\tan \varphi' = \tan d \sec t$

$$\cot Z = \frac{\sin (\varphi' - L) \cot t}{\cos \varphi'} \qquad (1)$$

(See Coffin's Prob. 38.)

$$\sin h = \frac{\cos (\varphi' - L) \sin d}{\sin \varphi'}$$

Should the local time not be well determined, it will be necessary to measure the altitude of the heavenly body at the same time that its horizontal angular distance from the base line is measured, either by the vertical circle of the theodolite, or by a second observer with sextant and artificial horizon. Then the body's azimuth will be found by either of the formulæ.

$$\cos \tfrac{1}{2} Z = \sqrt{\cos s' \cos (s' - p) \sec L \sec h} \quad \text{in which} \qquad (2)$$
$$s' = \tfrac{1}{2} (L + h + p), \text{ or}$$

$$\text{Sin } \tfrac{1}{2} Z = \sqrt{\cos s'' \sin (s'' - a) \sec L \sec h} \quad \text{in which} \qquad (3)$$
$$s'' = \tfrac{1}{2} (\text{co } L + d + h)$$

(See Coffin's Prob. 40.)

The correction for sun's semi-diameter when angle is measured by theodolite is $S' = S \sec h$

22. Should a Theodolite not be available the sextant may be used.

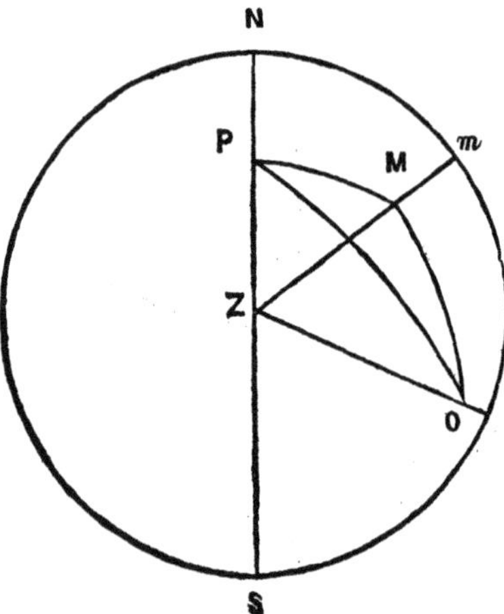

It will be at once seen that the angular distance thus measured is in the plane passing through the two objects, whereas when the theodolite was used the distance was measured in a horizontal plane. When the sextant has been used it becomes necessary to reduce the measured distance to a horizontal plane, or in other words to find the arc of the horizon intercepted between the vertical circles passing through the signal and heavenly body: that is, the difference of their azimuths. This may be done by the following formula. (See Coffin, Prob. 62.)

$$\sin \tfrac{1}{2} \zeta = \left(\frac{\sin \frac{1}{2} (D + H' - h') \sin \frac{1}{2} (D - H' + h')}{\cos H' \cos h'} \right)^{\frac{1}{2}}$$

let $d = H' - h'$

$$\therefore \sin \tfrac{1}{2} \zeta = \left(\frac{\sin \frac{1}{2} (D + d) \sin \frac{1}{2} (D - d)}{\cos H' \cos h'} \right)^{\frac{1}{2}} \qquad (4)$$

or
$$\cos \tfrac{1}{2} \zeta = \left(\frac{\cos \frac{1}{2} (H' + h' + D) \cos \frac{1}{2} (H' + h' - D)}{\cos H' \cos h'} \right)^{\frac{1}{2}}$$

let $S = \frac{1}{2} (H' + h' + D)$

$$\therefore \cos \tfrac{1}{2} \zeta = \left(\frac{\cos S \cos (S - D)}{\cos H' \cos h'} \right)^{\frac{1}{2}} \qquad (5)$$

M = body. O = Terrestrial object. H' = app. alt. of body. h' = app. alt. of terrestrial object. D = angular distance between two, corrected for index correction and semi-diameter ζ = diff. of azimuth between object and body.

If the object is in true horizon (or its alt. = Dip + Ref) then $\cos \zeta = \cos D \sec H'$.

Should the sun's limb (when angular distance is measured by a sextant) be brought tangent to a vertical edge of a terrestrial object (as the signal of a station), the correction for semi-diameter then becomes $s' = s \sin MOZ$.

MOZ, the angle between the vertical circle passing through terrestrial object and the plane passing through this object and the sun, may be found by $\cos MOZ = \dfrac{\sin H'}{\sin D'}$

D′ being measured distance corrected for index correction. It is much better, however, to measure the least distance to a point of the terrestrial object, and to this distance apply the whole semi-diameter and index correction, being careful that the altitude of the same point is observed. Making the sun's limb tangent to the vertical line is a difficult operation.

Then azimuth of terrestrial object (or base line) equals azimuth of heavenly body $\pm \zeta$.

23. For the purpose of referring azimuths, observed at night, to the direction of any geodetic signal, a mark is set up, consisting of a perforated box (about $\frac{3}{4}$ foot cube) through the front face of which the light of a bull's eye lantern is shown, appearing about the size and brilliancy of the star observed. The distance of this mark from the station is generally determined by local circumstances; but should, if possible, not be nearer than a mile.

If it is desired to obtain the azimuth of the same point by day observations, a vertical black stripe is painted on a wand, placed directly above and below the centre of the aperture, and of same width. If this is a quarter of an inch in width it will subtend at the distance of one mile 0.″8.

24. Having thus determined the azimuth of any line and wishing to set up a mark which will show the direction of the meridian passing through the spot at which the instrument is placed, it is only necessary to turn the telescope (pointed at the object whose azimuth has been determined) through an angle equal to the azimuth, and in the opposite direction to that in which the azimuth was reckoned.

If now a stake be set up in the prolongation of the line of sight, as shown by the cross-wires of the instrument, then the line drawn from this object to the centre of the station occupied will be the *meridian line.*

DETERMINING THE MERIDIAN LINE BY SHADOWS.

25. A good approximation may be made as follows: Plant a stake upon a level piece of ground, and give it a vertical position by means of a plumb line. Describe one or more circles on the ground from the foot of the stake as a centre. At the two instants before and after noon, when the shadow of the stake extends to the same circle, the azimuth of the shadow east and west are equal. The points of the circle at which the shadow terminates at these instants being marked, let the included arc be bisected; the point of intersection and the centre of the stake then determine the meridian line. Theoretically, a small correction should be made for the sun's change of declination, but it would be quite superfluous in this method. (Chauvenet's Astronomy. Vol. I.)

26. As this text is to be used by those who, in course of instruction in navigation, have already learned the various methods of rating chronometers, and of determining the Latitude and Longitude, the formulæ and general discussions will not be touched upon here. They may be found in Coffin's Navigation, or still more elaborated in Chauvenet's Astronomy.

In order that the different points of the triangulation may be determined in Latitude and Longitude, it is necessary to fix accurately the Latitude and Longitude of one of the principal stations; then by computation the Latitude and Longitude of the stations may be found; or, after the work has been plotted, they may be taken directly from the chart. The station of which the position is to be thus fixed by observation is usually one of the base stations.

The Latitude should, when possible, be fixed by observation with the zenith telescope, and the Longitude determined by means of the transit instrument and chronometer. These instruments, not being available, the sextant and artificial horizon are used.

The Latitude may be determined by observations of the sun or fixed stars when on the meridian, or by circum-meridian altitudes of those bodies. Stars are the more generally used; for in a few hours one is able to observe several different stars, whereas the sun is in proper position for the determination of the Latitude by these methods but for a few minutes during the twenty-four hours. In employing the method of circum-meridian altitudes of fixed stars it is best to choose those which cross at nearly the same time, and at nearly equal zenith distances, but on opposite sides of the zenith.

The Longitude must be determined by the ordinary chronometer sights.

CHAPTER III.

TRIANGULATION.

THE SELECTION OF STATIONS—SIGNALS—MEASUREMENT OF HORIZONTAL ANGLES—REPEATING ANGLES—FORM OF ANGLE BOOK—REDUCTION TO THE CENTRE OF STATION—REDUCTION OF SEXTANT ANGLES TO THE HORIZONTAL PLANE—SPHERICAL EXCESS—THREE ORDERS OF TRIANGULATION.

SELECTION OF STATIONS.

27. In the selection of principal stations the following points should be kept in mind:

1. The number of stations will depend upon the amount of detail to be shown upon the chart—stations should be placed upon all important topographic and hydrographic points.

2. They should be chosen so that the angles of the triangle may be as nearly equal as possible; no angle to be less than 30°.

3. The length of side should not exceed the power of the telescope of instrument.

4. At the end of the triangulation the station should be chosen so that a verification base may be measured.

It frequently happens that the only base line that can be taken lies in such a direction as to give poor triangles for the determination of other important stations. When this is the case it is necessary to choose some station which can be accurately fixed from the base line, and then use the computed sides of this triangle as the base in the computations neccessary to determine the position of those points which could not be determined by triangles formed upon the original base.

NAMES OF STATIONS.

The station should be called after the popular designation of the site on which it is situated, or after some peculiarity of the ground or formation well known in the neighborhood. The same name should not occur twice in the same section of the country. Or the stations may be numbered or distinguished by the letter of the alphabet. (Coast Survey Report.)

SIGNALS.

28. As it rarely occurs that natural objects are suitable or properly placed to indicate the angular points of the triangulation, and as such objects as light-houses, towers, &c., on the centre of which the instruments cannot be placed (thus rendering certain reductions neccessary), are inconvenient for use, it generally becomes necessary to erect signals at these points. The signal should always be so arranged that the theodolite may be placed directly over or under the center of the station, and the signal pole, if removed, exactly replaced. It is also necessary that the height of signal be sufficient to overcome the difference of level due to curvature.

The signal ordinarily consists of a pole from 10 to 25 feet in length placed vertically and surmounted by a flag, and steadied by braces. (For a variety of signals, see Coast Survey Reports or Jeffer's Surveying, pages 24, 25.) In regard to the diameter of the pole to be used, it is evident that, in short lines, it should not exceed the size just sufficient to admit of its being distinctly seen, as all increase of size is a source of error. With sides not exceeding 5 miles, pole not to exceed two to five inches in diameter; with sides upward of 5 miles, from 5 to 8 inches (Coast Survey Report). In the erection of any kind of signal too much care cannot be used in centering the mark to be observed exactly over the center of the station.

29. In the primary triangulation, in which the length of the sides is very considerable, it is recommended in the Coast Survey Reports that the line of sight of the telescope should never be allowed to pass within less than 6 feet from the ground.

In order to determine the required height of signal of any given length of line, allowance being made for terrestrial refraction and curvature, the tables on page 54 may be referred to.

As an illustration of the use of this table, let us suppose that a line, A to B, was 18 miles in length over a level plain, and that the instrument could be elevated at either station, by means of a portable tripod, to a height of 20, or 30, or 50 feet. If we determine upon 36.3 feet at A, the tangent would strike the curve at the distance represented by that height in the table, viz., 8 miles, leaving the curvative (decreased by the ordinary refraction) of 10 miles to be overcome. Opposite to 10 miles we find 56.7 feet, and a signal at that height erected at B would, under favorable refraction, be just visible from the top of the tripod at A, or be on the same apparent level. If we now add 8 feet to tripod and 8 feet to signal pole, the visual ray would certainly pass 6 feet above the tangent point, and 20 feet of the pole would be visible from A. Other variations may be made to suit the circumstances, as it rarely happens that in lines across water or level land there are not some slight elevations which could be made available as stations.

CONSTRUCTION OF SIGNALS.

The following instructions in case of large triangles is taken from Coast Survey Report. 1868.

In no case should the foot of the pole be inserted in or be allowed to come in contact with the earthenware cone, or other article buried as an underground mark. Six or eight inches of earth carefully packed above the cone or block, and upon this a square foot or so of board, upon which the pole can rest, will be sufficient to afford a foothold when such is necessary, and, at the same time, to prevent any displacement of or injury to the mark, in case the pole should be roughly handled or blown down.

The tripod and scaffold, which are frequently erected for the elevation of the instrument and observer, in order to obtain a longer length of line or to escape the troubled condition of the atmosphere usually lying immediately over the low, flat lands bordering the coasts and shores, vary in height from 10 to 60 feet, and are made of scantling, purchased for the purpose, or of materials obtained from the forest, and are built in general accordance with the plans and specifications given in the annual reports. While a strict adherence to uniformity in the details of the construction is not possible, so much depending upon the means and facilities at the disposal of the assistant and upon a proper regard to economy, the general principles of strength and solidity, in both tripod and scaffold, are strictly observed, by a proper spread and anchoring of the feet, a thorough bracing of the legs, and a compact fitting of the cap to the top of the tripod. A careful attention to these points will secure perfect immobility while the observations are being made, and sufficient firmness to keep the scaffold entirely and always free from contact with the tripod, and to enable both structures to resist the most violent storms.

The following description and plan of a small portable tripod and scaffold may be of service, the height being well adapted for short lines on the southern coast:

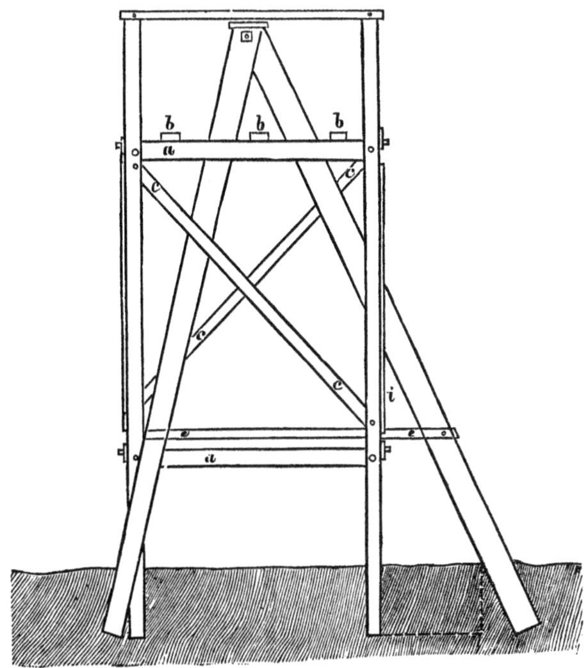

Tripod.—This is made of three legs 18 feet long and six inches in diameter, bolted together, 3 inches from the top, with an inch bolt, 16 inches long. After they are fitted together at the head, the spread being 13 feet, and before raising, one of the braces, *ee*, is screwed to the two outside legs so as to keep them in place while raising the tripod by the third leg. It is then settled two feet in the ground, care being taken to level it by the braces *ee*, which are to be screwed on with wood screws.

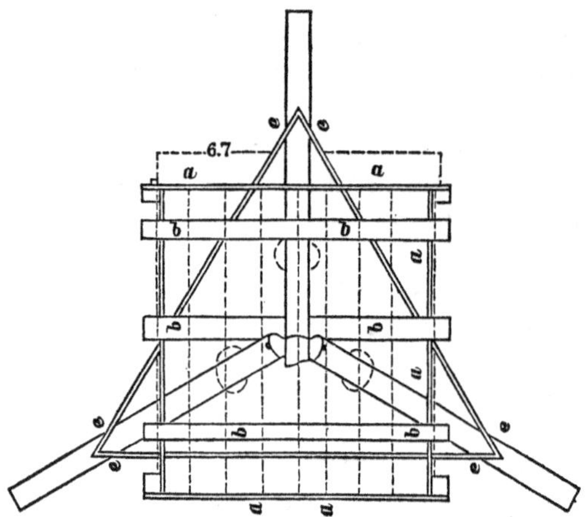

Scaffold.—4 posts, 16½ feet long and 5 inches square.
8 cross-pieces, 7 feet long and 6 inches by 1¼, *a a*.
3 cross-pieces, 7 feet long and 6 inches by 1, *b b*.
4 braces, 10 feet long and 5 inches by 1¼, *c c*.
8 flooring pieces, 7 feet 3 inches long and 9¼ by 1, with holes for the tripod legs.

The end posts are those on which the braces and cross-pieces screw on the outside, and which are to be fastened together in pairs, when on the ground, so that it may be raised after the manner of a bedstead or house frame. The braces, *c c*, are to be bolted to the upper ends to steady the posts when raising them. All the holes in the posts, cross-pieces, and braces are to be identical as to plan and size, as also the pieces themselves, so as to have no mistakes, and when raised the scaffold fits to the tripod as shown in the drawing. The scaffold is levelled by the cross-pieces and adjusted, and in firm position after the floor is on, so as to be free from the tripod. The floor is to be 2 feet 10 inches below the top of the tripod. Three iron knees are screwed to the tripod legs near the top, so that the triangular piece for the theodolite can be bolted to them. This piece is made of two pieces of one-inch plank screwed together across the grain, and then painted. Holes are made in the floor for the tent posts, and wire guys are sometimes required for the scaffold.

The above takes about two hours to put up. One large and two small wrenches are necessary; also a bag to contain the screws, nuts, &c. Should more than one be needed, they should be painted different colors, but be in all other respects exactly alike, in order that one can be used to repair the other.

The station marks include the underground and surface marks; the former to be buried and the latter to be thrown up for the preservation of the centre and of the position of the station.

MEASUREMENT OF HORIZONTAL ANGLES.

30. When the theodolite is used, the following is the method pursued.

Set the theodolite over the station (from which angles are to be measured) so that the plumb bob shall be exactly over the centre of the station, stretching the legs wide apart, and placing them in solid ground so that they may remain firm during the observations. By a little practice the above may be carried out, and the azimuth plates at the same time be laid nearly horizontal. By means of the levelling screws and the levels bring these plates horizontal, turning them around slowly to verify the adjustment. Then place the 0° of upper plate in coincidence with 0° of lower plate and clamp it in that position. Unclamp lower plate and turn the telescope toward the station which has been chosen as the origin of angles. By means of the tangent screw attached to lower plate, the cross-wires are made to bisect the signal of the station (call A). Now unclamp the upper plate and turn the telescope to the right, clamp, and by means of the tangent screw attached to the upper plate, bisect the signal of station (call B). whose angular distance from origin (A) it is desired to measure, leaving the upper plate securely clamped, and take the reading of both verniers. Now unclamp the lower plate and turn it to the left till the telescope again points to the signal of A, when the lower plate is again clamped and, by means of its tangent screw, the cross-wires are made to bisect the signal. Unclamp upper plate and again turn telescope in same manner as before to bisect the signal of B, and again read both verniers. This reading should be exactly the double of the proceeding one; for after having *once* measured the angle, the 0° of the graduated (lower) plate has been moved to the left as much as the telescope had been before moved to the right; hence when the telescope is the second time moved to the right to bisect the signal B, the reading should be double the angle measured. Continue in this manner until at least a round of the circumference of the plate has been passed over. This method is called "repeating" the angles; its object is to lessen the errors of observation and to eliminate as much as possible the errors of graduation and the eccentricity of the circle. The final reading is taken, and if more than the circumference has been passed over, 360° is added to it, and the sum divided by the number of times the angle has been repeated will give the angle accurately. It is recommended to repeat the angle six times and to pass over at least one circumference of the circle.

31. To measure a round of angles without "repeating" them. When it is considered a needless refinement to "repeat" the angles, as before described, a round of angles may be measured in this way.

Set the theodolite up over the centre of the station to be occupied and, after having levelled the instrument, place the two zeros (of the plates) in coincidence; clamp the upper plate, and turn the lower plate till the signal of the station selected as original of angles comes into the field of the telescope; then clamp the lower plate, and by means of its tangent screw bisect the signal by the cross-wires. Then unclamp upper plate and turn telescope to the station next to the right, clamp, and, by means of tangent screw of upper plate, bisect this signal by the cross-wires; then read both verniers. Unclamp upper plate (leaving lower plate still clamped) and turn to the next station to the right; perfect the bisection of this station with the cross-wires, and read as before. Proceed in this way until the entire round has been measured. Then turn the telescope to bisect the signal of origin again, and if the zeros still coincide, the instrument has not been moved. This is called "verifying the zero." If the coincidence of the zeros is not perfect the angles must be measured again.

The following tables are given showing the manner of recording the angles; the form being taken from the Coast Survey Angle Book. The quantities in the columns are taken from actual work, and are given as examples merely.

FORM FOR RECORDING HORIZONTAL ANGLES.

No.	Time.	Angles.	Objects observed.	Station occupied.	Remarks.
Feb. 20	9 A.M.	14.05.10	Station B	East Base	Origin of Angles.
		28.10.30	......	......	West Base.
		42.15.20	......	......	Angles repeated six times, both verniers read, and same result obtained.
		56.20.40	......	......	Instrument used.
		70.25.30	......	......	Wurdeman Theodolite —8″ circle.
		84.31	......	......	Least count of vernier 10″
		6)84.31	......	......	Weather overcast.
		14.05.10	......	......	
	9.50	69.43.20	Station E	do	Same origin.
		139.26·40			Same remarks.
		209.10.10			
		278.53.40			
		348.36.20			
		58.20.00			
		360			
		6)418.20.00			
		69.43.20			

SHOWING A ROUND OF ANGLES.

No.	Time.	Angles.	Objects observed.	Station occupied.	Remarks.
Feb. 21	9.40 AM	20.05	E	C	Origin D.
		42.08	F	......	Instrument used.
		55.07	G	......	Bureau Theodolite.
		87·34	EB	......	Least count, 1′ 00″
		102.16	WB	......	A and B not in sight.
		0°-1′	D		

REDUCTION TO CENTRE OF STATION.

32. Frequently it is not possible to occupy the centre of one of the stations chosen, so that the angles, measured at some place near the centre of the station, must be reduced to what they would have been had they been observed at the centre of the station. Such correction or reduction is called the "Reduction to the Centre of Station."

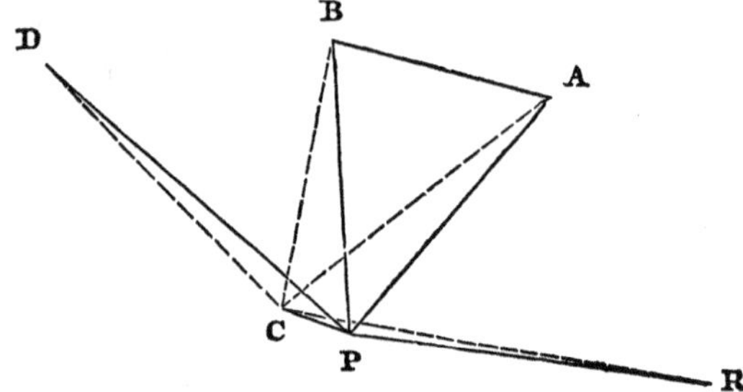

In figure let C = centre of station.

P = Place occupied by instrument.
O = Angle at P, between the stations A and B.
y = Angle at P, between C and left-hand station B.
r = Distance CP.
D = " AC.
G = " BC.

Call the angle at C between A and B = C.

In triangle CBA, $C = 180° - CBA - CAB =$ (1)
$180° - (CBP + PBA) - (BAP - PAC)$

In triangle CAP, $\sin(O + y) : \sin PAC = D : r$

$$\sin PAC = \frac{r \sin(O + y)}{D} \qquad (2)$$

since PAC is small, $PAC = \dfrac{r \sin(O + y)}{D \sin 1''}$

In triangle CBP, $\sin y : \sin CBP = G : r$

$$\sin CBP = \frac{r \sin y}{G}, \quad CBP = \frac{r \sin y}{G \sin 1''} \qquad (3)$$

$$C = 180° - PBA - BAP - \frac{r \sin y}{G \sin 1''} + \frac{r \sin(O + y)}{D \sin 1''} \qquad (4)$$

But 180° – PBA – BAP = O

$$\therefore C = O - \frac{r \sin y}{G \sin 1''} + \frac{r \sin (O + y)}{D \sin 1''} \qquad (5)$$

EXAMPLE OF REDUCTION TO CENTRE OF STATION.

Taken from Jeffers' Surveying.

G = 42390 yards	r = 5.00 yards	D = 57836 yards
O = 42° 53′ 21	y = 248° 50′	O + y = 291° 43′ 21″

Log r =	0.69897		
ar. co. log. sin. 1″ =	5.31443		
log. $\frac{r}{\sin. 1''}$ =	6.01340	. . . .	6.01340
log. sin. (O + y) =	–9.96801	log. sin. y =	– 9.96966
ar. co. log. D =	5.23781	a r. co. log. G =	5.37273
2d Term–16.″57	log.–1.21922	1st term–22″. 69	log,–1.35579

1st do. +22.″69

+6″.12 = Reduction to centre.

REDUCTION OF SEXTANT ANGLES TO HORIZONTAL PLANES.

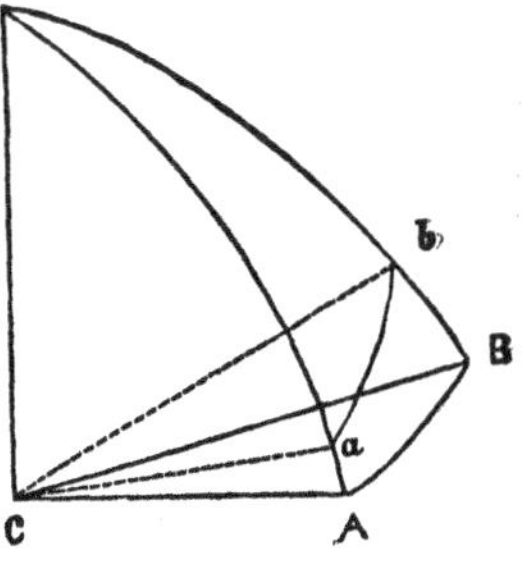

If the angular distance ab has been measured in an oblique plane, it becomes necessary to find from this the horizontal angular distance between the vertical circles passing through the two points; that is, to obtain the horizontal angle between them. If we measure the altitudes of a and b, and thence get their zenith distances, we have in the Spherical Triange a Zb, the sides Za, Zb, and ab, from which can be found the angle aZb, which is measured by the arc AB, which also measures the angle AcB, which is the horizontal angle between the two points a and b.

To find this the following formula may be employed.

$$\sin \tfrac{1}{2} \zeta = \left(\frac{\sin \tfrac{1}{2} (D + H - h) \sin \tfrac{1}{2} (D - H + h)}{\cos H \cos h} \right)^{\frac{1}{2}} \qquad (1)$$

putting $d = H - h$

$$\therefore \sin \tfrac{1}{2} \zeta = \left(\frac{\sin \tfrac{1}{2} (D + d) \sin \tfrac{1}{2} (D - d)}{\cos H \cos h} \right)^{\frac{1}{2}} \qquad (2)$$

in which H = alt. of a, h = alt. of b, D = ang dis. cor. for index correction, ζ = ang aZb.

$$\text{or} \cos \tfrac{1}{2} \zeta = \left(\frac{\cos \tfrac{1}{2} (H + h + D) \cos \tfrac{1}{2} (H + h - D)}{\cos H \cos h} \right)^{\frac{1}{2}} \qquad (3)$$

which are the same as found in the discussion of the astronomical bearing.

These formulæ are general and exact in all cases: but as the altitude of a and b rarely exceeds a few degrees, it is usual (when great accuracy is desired) to compute the correction of the observed angular distance to obtain the corresponding horizontal distance, the formulæ for which may be found in Jeffers' Surveying, page 131. But as in the following example, in which $h = 2°$, $H = 3°$, and the distance measured 56°, it was found that the correction amounted to but 0′ 50″, it will be seen (bearing in mind that it is practically impossible to measure angles with the sextant within 20″) that this correction can be generally neglected, provided the altitude of neither object exceeds 2° or 3°.

If the plane in which the objects lie is inclined at a small angle to the horizon, a very good value of the horizontal angle may be obtained by observing the angular distance of each object from a third, very far to the right or left of both. The difference of these angular distances will be nearly equal to the desired horizontal angle. Should the obliquity of the plane be considerable, the horizontal angle must be found by the formulæ given.

SPHERICAL EXCESS.

34. Definition.—The excess of the sum of the angles of a spherical triangle over two right angles is called the spherical excess.

In small triangles where the sides do not exceed 6 or 8 miles it need not be considered, but in large triangles it must be.

The expression for the spherical excess is [Chauvenet, Trig. Art. 162.]

$$\text{Excess} = \frac{a.\ b.\ \sin C}{2} \cdot \frac{1}{r^2 \sin 1''}$$

$$\frac{a.\ b.\ \sin C}{2} = \text{area of triangle.}$$

$$r = \text{radius of earth.}$$

logarithm of mean radius of earth in yards is 6.8427917

Between latitudes of 25° and 45° the spherical excess amounts to about 1″ for an area of 75.5 square miles. Hence if the area in square miles be known, a close approximation to the spherical excess will be had by dividing the area by 75.5 miles.

If the three angles are assumed to have been equally well measured, the previous determination of the spherical excess is not necessary for the calculation of the sides, (though it will be required for estimating the relative accuracy of the observations), for the length of the sides of a spherical triangle may be computed as if they were rectilinear when one third of the excess of the sum of the three angles above 180° is deducted from each of the three observed angles.

$$b = \frac{a \sin (B - \frac{1}{3} E)}{\sin (A - \frac{1}{3} E)}$$

$$\text{For large triangles excess} = \frac{a.\ b.\ \sin C\ (1 + e^2 \cos 2\ L)}{2\ r^2 \sin 1''}$$

in which latter r = equatorial radius, and L = the mean latitude of the three stations.

$$e = \text{eccentricity of the earth.}$$

35. Three orders of triangulation are recognized in the geodetic operations for the survey of a coast.

1. The primary triangulation with sides varying from 20 to 100 miles in length.
2. The secondary triangulation with sides varying from 5 to 20 miles in length, either connected with the primary, or starting out from independent bases.
3. The tertiary triangulation with sides of less than 5 miles.

The survey of a harbor therefore generally comes under the head of tertiary triangulation.

To unite the most advantageous conditions, as has been before stated, the triangles should be as nearly equilateral as possible. Small errors in the measurement of angles will then least affect the computed length of opposite sides. The least number of triangles which will cover the ground should be employed; but for the purpose of a hydrographic survey the sides should seldom exceed 12 or 15 miles, as it would be difficult to distinguish the signals in the atmosphere of the sea-shore.

On the coast survey, with very powerful instruments, triangles with sides of 40 miles are common, and sides of 84 miles have been used.

A verification base should be measured and connected with one of the last triangles, to detect any errors which may have accumulated in the triangulation.

CHAPTER IV.

PLOTTING THE WORK.

THE SCALE—PLOTTING THE PRINCIPAL STATIONS—RUNNING IN THE SHORE LINE—PLOTTING IT—THE TELEMETER.

THE SCALE OF THE CHART.

36. The scale of the chart will depend upon the size of the paper and the extent of the survey. Scales are usually expressed fractionally.

$$\frac{1}{5{,}000} \qquad \frac{1}{50{,}000} \qquad \frac{1}{100{,}000} \text{ \&c.}$$

Wishing to construct a scale of $\frac{1}{5000}$ lay off on the edge of the paper a line one, two, or any number of metres or feet in length. Divide each metre or foot into fifty equal parts. Each foot will represent 5000 feet, and each subdivision 100 feet. Lay off an additional subdivision (100 feet) to the left of the zero of the scale, and divide it into five or ten equal parts, and upon this construct a rectangle and draw a diagonal scale.

37. The angles having been measured at the different stations, and the triangles having been determined upon, the sum of the angles at the three stations is taken, which (in the triangle of an ordinary harbor survey) should equal 180°. Should the sum not equal 180°, and should equal reliability be placed upon each of the angles, then one-third of the difference of the sum and 180° should be added to, or subtracted from, each of the angles.

Now having the angles and the length of the base line given, find the length of the other two sides. These sides may now be used as the given sides of other triangles, but whenever it is possible the triangles should be formed upon the base line, so that the errors may be carried through the work as little as possible.

Before plotting the stations upon the chart it is necessary to plot the base line in its proper direction and length.

Draw a line to represent the meridian (usually drawn parallel to one edge of the paper), and from any point on it lay off the azimuth of the base line; draw a line in this direction, and in such a part of the paper that all the stations may be plotted. Take from the scale the length of the base, lay it off on the line which has been drawn as the projection of the base line produced, and the extremities of the portion laid off will be the base stations.

To plot the other stations it is best to compute the sides as described, and with the length of the side from Base (1) to A taken from the scale as a radius and Base (1) as a centre, describe an arc; with the length of the side from Base (2) to A taken from the scale and with Base (2) as a centre describe a second arc,—the intersection of the two is the position of the station A. In this manner all the stations may be plotted.

This is more accurate than by laying off the angles from Base (1) and Base (2), and fixing A by the intersection of the two lines Base (1) A, Base (2) A; but in reconnoissances, and with the not very precise instruments for measuring the angles with which surveyors are sometimes supplied, the points will probably be fixed with sufficient accuracy by intersection. (It will be seen that perfect rulers and protractors are necessary in this method particularly if distances are considerable.) Other prominent points may be plotted upon the chart by "cuts;" measuring at two or more principal stations the horizontal angles between the prominent points (which may be called secondary stations) and some other principal station which has been accurately placed upon the chart by the computed lengths of the sides of the triangle by which its position was fixed. These secondary stations, which do not enter at all into the main system of triangles, are chosen entirely with reference to their topographical or hydrographical importance. In this work, the minute accuracy necessary for the determination of the principal stations is not required. Fixing several secondary stations about the shore or coast in this way affords a good check upon the coast line between the principal stations.

TO RUN IN THE SHORE LINE BETWEEN TWO PRINCIPAL STATIONS

38. This may be accurately done in the following manner, using a theodolite and a surveyor's chain, steel tape, or a line carefully marked and compared with some standard measure. Set the theodolite up at one of the principal stations, place the zeros in coincidence, clamp the upper plate and point the telescope at some other principal station, which should be so chosen that lines drawn from it to parts of the line to be run shall not make too acute or too obtuse an angle; clamp lower plate. Send a man with a flag along the line, who is to proceed as far as the contour of the coast or the character of the country will allow it to run without changing its direction, and let him place the flag at this point (*a*). Turn the telescope of the theodolite to bisect the flag pole; clamp upper plate and note the reading. The measuring party now move along the line, being kept in it by the observer at the theodolite. The leading chain-man carries in his hand ten pins. The rear chain-man places his end of the chain at the centre of the station started from and the leading chain-man stretches the chain taut along the line, and sticks a pin in the ground at the end of the chain. Both chain-men now move on till the rear man comes up to the pin stuck in the ground. When the rear end of the chain is brought to the pin and the chain again hauled taut in the line and the head chain-man sticks a second pin in the ground. The rear man now pulls up the first pin, and both move on. This is repeated till all the pins have been stuck by head chain-man; when the rear chain-man pulls up the tenth pin an entry is made in the record book and the pins are sent on to head man. At every length of the chain, or more or less frequently, according to the nature of the coast or country, offsets are measured at right angles to the line, to the water's edge, or the hills, or any prominent landmark. The distance having been measured, the theodolite is moved forward to the auxiliary station (*a*) marked by the flag pole, and is set up. The lower plate is unclamped and the telescope turned toward the station just left—the lower plate is then clamped and the telescope transited, the upper plate unclamped and the telescope pointed in the new direction of the line, indicated by the flagman who has been sent on to the point (*b*) where the line again changes its direction: the reading is then noted. The chaining party proceeds as before, and, having

measured this second distance, the theodolite is again moved forward; but the place last occupied by it must be marked by a flag, or some other distinguishable signal, so that when the theodolite is set up at a third point (*c*) it may be sighted back to (*b*). This process is continued till one of the principal or secondary stations is reached, when this station becomes the origin of future measurements.

The coast line is then *plotted* upon the chart in the following manner: (Fig. A). Draw a line from station left to the origin of angles. Lay off the angles which the different portions of the line make with the line drawn from station left to origin. Lay off in the first direction the first distance measured and mark the point (*a*). With a pair of parallel rulers draw a line passing through (*b*) parallel to the second direction of the line, and lay off the distance measured between (*a*) and (*b*) mark this point (*b*), and so on till all points are plotted. The offsets are laid off, in their proper intervals, at right angles to the measured line, and a free hand line is drawn through the ends of the offsets which will represent the shore line; or if offsets have also been measured back to the hills, it will give their contour.

In the figure the principal station started from is F, and the station chosen as the origin of angles is A. The field notes may be kept as in (Fig. B).

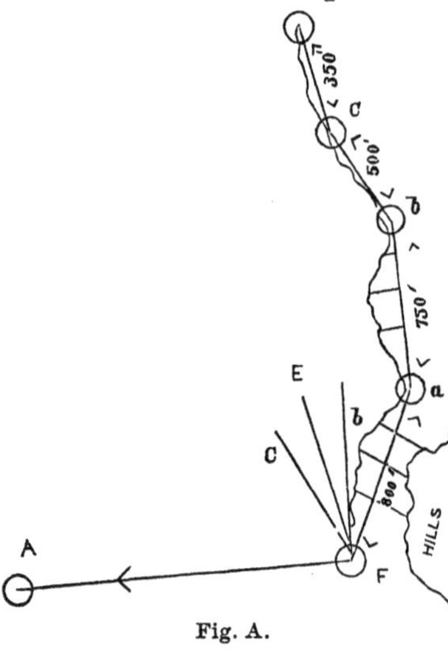

Fig. A.

By this method of running in and plotting the coast line it will be noticed that all angles are laid off at the same point and from the same origin; consequently the only error that will enter in the plotting will be in the lengths of the measured lines (supposing the angles to have been correctly measured), for the directions of the lines will be same as they are in reality. It will be seen that less error is accumulated by this method than by the others hereafter described.

It will be found that wharves and *streets* may be put in in this manner very accurately.

39. To run the shore line in with Bureau of Navigation pattern of theodolite or sextant.

The theodolite furnished by the Bureau of Navigation has but one horizontal plate which is capable of motion around a vertical axis; consequently it is impossible to repeat angles, and it is also impossible to run in a shore line by the method (using back angles) before described. To accomplish the desired result with this instrument, set it up over the principal station to be started from, and having adjusted the instrument turn to one of the other principal stations, which lies in a direction as near 90° as possible from the direction of line to be run, and note the reading of the zero. Having sent a flagman ahead, as far as the contour of the coast or the nature of the land will allow, unclamp the plate and turn the telescope so that its cross-wires bisect the flagman's signal, and note the reading.

The measuring party now proceed to measure the distance between the principal station and the subsidiary station (*a*), as before described, being kept in line by an observer at theodolite. This done the theodolite is removed and (the signal having been replaced) is taken to the subsidiary station occupied by flagman, who now moves ahead to the next turn of the coast (*b*). The theodolite is to be set up at (*a*), and the telescope is turned to one of the principal stations, not necessarily the same one as before observed; the angle between this station and the new direction of the line observed, and the distance between (*a*) and (*b*) measured. Thus proceed until the other principal station is reached.

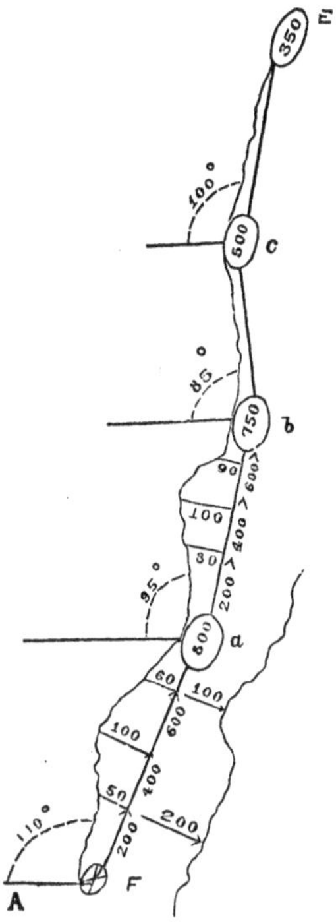

Fig. B.

The method of plotting this work is of course different from that employed in the first method described; and it will be seen that in using this method an error in any measured distance will affect the result much more than the same error would by the first.

The work is plotted in this way. Suppose the start to be made at station C. The angle the first direction of the line makes with a

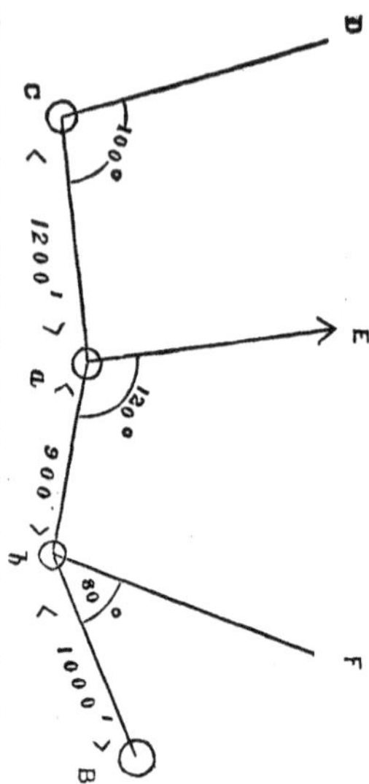

line from C to D is 100°; its length is 1200 feet. The angle that the *second* direction of line makes with a line drawn from (*a*) to E is 120°; its length 900 ft. The third direction, from (*b*) to station B, makes an angle of 80° with a line from (*b*) to F, its length 1000 ft. At C lay off a line making angle of 100° with C D, and on the line mark off 1200 ft. according to scale; this will give (*a*). At (*a*) lay off a line making an angle of 120° with the line (*a*) E, and mark off the distance as before. So proceed till station B is reached. If the end of the line (*b*) B falls upon B the work is correct.

But it will be seen that should the distance from C to (*a*) be in error, that not only will the position (*a*) be in error, but the plotted direction of the line from (*a*) to (*b*) will also be in error.

The figure will also serve as a model for keeping the field notes. There are, however, no offsets shown in it.

See plate IX. in reference to scale, triangulations and plotting of shore line.

The Telemeter.

40. In consequence of some of the disadvantages resulting from the employment of the chain (or measured lines), among which are the necessity of dependence for correct distance upon the chain-men, the number of persons required, the time consumed, and the obstacles to its use in some parts of the coast line or country, an instrument called the *telemeter* may be used. Experience has shown that distances measured by it have been exceedingly correct.

The telemeter, as used by the Coast Survey, is simply a scale of equal parts painted upon a wooden plank about 10 feet long, 5 inches wide, and $1\frac{1}{4}$ inches thick; so graduated and marked that the number of divisions upon it as seen between the upper and lower wires (horizontal) of the telescope of a theodolite or plane table is equal to the number of units in the distance between the observer's eye and telemeter held at right angles to the line of sight. In all cases the telemeter should be graduated experimentally for the particular telescope and eye of the observer who is to use it. Although the telemeter is intended for use in connection with the plane table, it can be also used with the theodolite by placing an additional horizontal wire in the diaphragm plate. The following is from the C. S. Report of 1865:

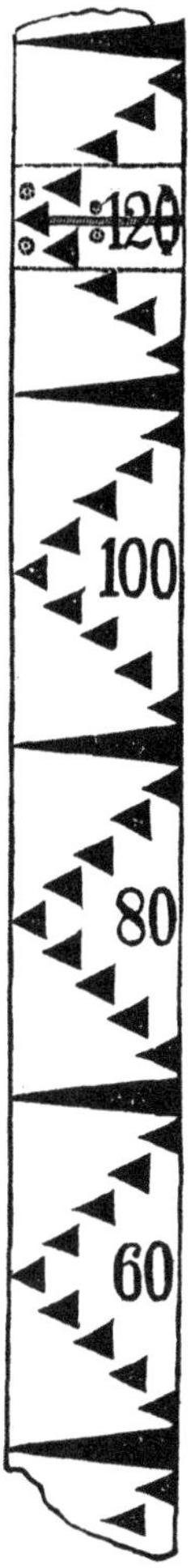

The horizontal wires in the diaphragm of the telescope should be accurately adjusted, and the divisions of the telemeter made to correspond in length with the distance included between the upper and lower wires of the telescope at a carefully measured distance, and then divided into as many equal parts as there are units in the distance measured.

For convenience of transportation it can be hinged in the middle, and secured on the side when in use by a sliding bolt; and as it is necessary, when observed upon, that it should be held accurately at right angles to the line of sight, a small brass movable bar, with sights or a groove upon its upper edge, should be fixed upon the side of the rod at a convenient height for the eye of the rodman, and which, when in position, will be perpendicular to the plane of the telemeter and directly in the line of sight of the telescope.

The correctness of the telemeter depends upon the closeness of the reading, and the accuracy with which the rod is held perpendicularly to the line of sight.

With ordinary care an error of reading should not occur even at the greatest distance denoted on the rod. With the observations carefully made, and the reading of the rod reduced to a horizontal plane, the greatest distance given by it—as usually divided—can be relied on as practically correct. There is no sensible error at any distance greater than 20 meters and less than 260, and, generally speaking, the telescopes of the Coast Survey alidades have not sufficient reading power beyond 400 meters, but it will generally be safe to rely upon it for any distance from 15 to 300 meters, beyond which it cannot be read with accuracy for use in constructing a map on a scale of $\frac{1}{10000}$.

The telemeter has been recommended for use in a great variety of cases where it becomes necessary to determine distances, in such close filling in as the corners of streets, wharves, &c., determination of all classes of detail, in traverses, shore lines, and even the establishment of positions; but in the latter it is safe only to depend upon good intersections. It has been employed, however, in all manner of detail, and is preferred by some to the chain in all cases save in compactly built streets and on long lines, where the distances are so great that the telescope will not admit of the accurate reading of the rod; it is maintained by some that where only a single point is to be seen positions can be readily and accurately determined.

For the reduction of the hypothenuse to the base, the following table is given:

TABLE FOR REDUCTION OF HYPOTHENUSE TO BASE.

Angle.	Hypothenuse.				
	100 meters.	200 meters.	300 meters.	400 meters.	500 meters.
5°	99.62	199.24	298.86	398.48	498.10
10°	98.48	196.96	295.44	393 92	492.40
15°	96.59	193.19	289.78	386.37	482.96
20°	93.97	187.94	281.91	375.88	469.85
25°	90.63	181.26	271.89	362.52	453.15
30°	86.60	173.21	259.81	346.41	433.01
35°	81.92	163.83	245.75	327.66	409.58
40°	76.60	153.21	229.81	306 42	383.02
45°	70.71	141.42	212.13	282.84	353.55

CHAPTER V.

DETERMINATION OF HEIGHTS.

MEASUREMENT OF SMALL ALTITUDES AND DEPRESSIONS—MEASUREMENT OF SLOPES WITH SEXTANT AND ARTIFICIAL HORIZON—DETERMINATION OF HEIGHTS.

41. To measure very small altitudes or depressions with the sextant and artificial horizon.

When the object has a considerable altitude it is measured in the same manner as in measuring the altitude of a heavenly body with the same instruments. But when the altitude is small this becomes exceedingly inconvenient and difficult. The following method may be employed for determining small altitudes or depressions.

Stretch a string over the artificial horizon. Place the eye so that the string will cover its own image in the mercury. The eye and string are consequently in the same plane.

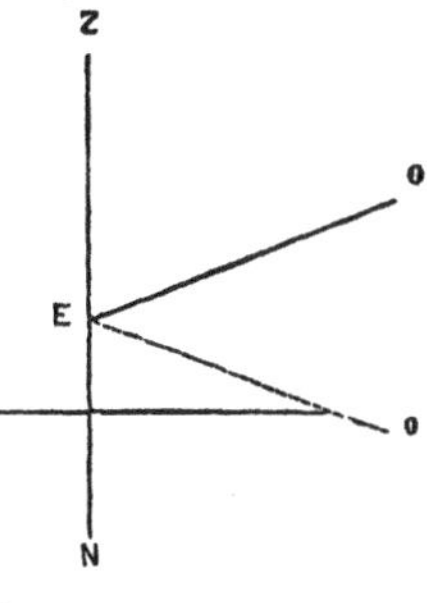

Then, seeing the string by direct vision, cause the reflection of the object, whose altitude or depression is to be measured, to coincide with the string. The angle measured is O E N, the supplement of the object's zenith distance, which, if greater than 90°, will give the object's altitude; if less, its depression.

42. To measure slopes with the sextant and artificial horizon.

Let A G be a horizontal line; and A B the surface or line whose slope is required. Mark two points equally distant from the place of observation, and on the line A B. Measure, by the method described in the preceding article, the angles a and a'. Then half the difference between these two angles will be the inclination of A B to the horizon.

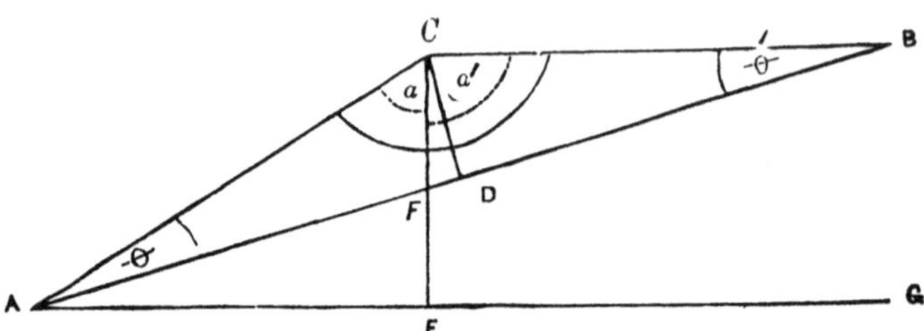

For the triangles A F E and C F D are similar, being right angled with their acute angles at F equal.

Consequently

$$DCF = DAE. \qquad \text{But}$$

$$DCF = \tfrac{1}{2}(a + a') - a = \tfrac{1}{2}(a' - a) =$$

angle of inclination of A B to A G, the horizon.

If the points A and B are not equally distant from the observer, but yet far apart, this method will still give a close approximation, the error, which is additive, being $\frac{1}{2}(\theta' - \theta)$, D not being in middle point of A B the triangle is no longer isosceles, but

$$180^\circ - (a' - DCF) = 90^\circ + \theta'$$

$$\therefore DCF = -90^\circ + a' + \theta'$$

$$180^\circ - (a + DCF) = 90^\circ + \theta$$

$$\therefore DCF = 90^\circ - a - \theta$$

$$\therefore DCF = \tfrac{1}{2}(a' - a) + \tfrac{1}{2}(\theta' - \theta)$$

DETERMINATION OF HEIGHTS.

43. While measuring a round of angles the altitude of prominent heights should be observed as well as their horizontal angular distance from other points of the triangulation. This may be easily done with a theodolite fitted with a vertical or altitude circle; but should the theodolite have no such arrangement, or should the sextant be the instrument used for the measurement of the horizontal angles, then the altitudes must be measured with the sextant and artificial horizon.

When the triangles are computed, the distance between the station at which the altitude is measured and the object itself is found; or if its position has been fixed by cuts, then the distance may be measured upon the chart.

Knowing the distance of the object and the angle which its height subtends, we can find the height by the solution of a plane right angled triangle.

But when the distance is at all considerable it becomes necessary to take into consideration the figure of the earth, and consequently to apply a correction to the computed height. Considering the figure to be that of a sphere (which will create an inappreciable error), points that *are* on the same *actual* level are not on the same *apparent* level, owing to the curvature of the surface.

44. Let the figure (which is of course very much distorted) represent a section of the earth. A, the point at which observer is stationed; B′, a point of which the vertical height above B is required. HAB′ is the altitude of B′ as seen from A. The triangle AHB′ differs but little from a right-angled triangle, and the error in B′H, from considering it such, will not be appreciable. But it is evident that to find the true height above B, it is necessary to add the correction BH, and this is called the correction for curvature.

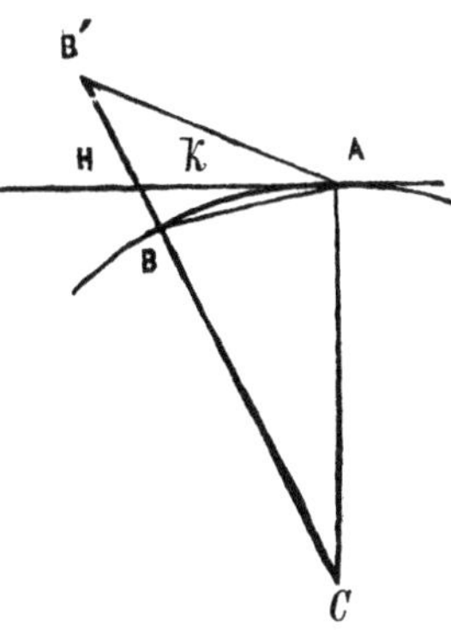

Let k = AH, known from computation or by measurement from chart. It is in reality the arc AB, but, being small, it will not differ sensibly from its tangent AH;

r = mean radius of earth;

x = correction for curvature.

Since C A H is a right-angled triangle

$$(x + r)^2 = r^2 + k^2$$

$$\frac{x^2}{2r} + x = \frac{k^2}{2r} \qquad (1)$$

but $\frac{x^2}{2r}$ will be so small a fraction that it may be neglected; at a distance of 24.3 statute miles $\underline{x}$ is about 300 ft.

$$\therefore \frac{x^2}{2r} = \frac{1}{450} \text{ ft. (about.)} \qquad (2)$$

These values for x are here tabulated.

DIFFERENCE IN FEET BETWEEN THE APPARENT AND TRUE LEVEL.

Distance, miles.	Difference in feet for—			Distance, miles.	Difference in feet for—		
	Curvature.	Refraction.	Curvature and refraction.		Curvature.	Refraction.	Curvature and refraction.
1	0.7	0.1	0.6	13	112.8	16.9	95.9
2	2.7	0.4	2.3	14	130.8	19.6	111.2
3	6.0	0.9	5.1	15	150.2	2 5	127.7
4	10.6	1.6	9.0	16	170.8	25.6	145.2
5	16.7	2.5	14.2	17	192.9	28.9	164.0
6	24.0	3.6	20.4	18	216.2	32.4	183.8
7	32.7	4.9	27.8	19	240.9	36.1	204.8
8	42.7	6.4	36.3	20	266.9	40.0	226.9
9	54.1	8.1	44.0	21	294.3	44.1	250.2
10	66.7	10.0	56.7	22	323.0	48.4	274.6
11	80.7	12.1	68.6	23	353.0	52.9	300.1
12	96.1	14.4	81.7	24	384.4	57.7	326.7

$$\text{Curvature} = \frac{\text{square of distance}}{\text{mean diameter of earth}}$$

$$\log.\ x = \log.\text{ square of distance in feet}—7.6209147$$

Refraction $= \frac{k^2}{R} m$, where k represents the distance, R the mean radius of the earth, and m the co-efficient of refraction.

$$\text{Curvature and refraction} = (1-2m)\ \frac{k^2}{2R}$$

Or, calling h the height in feet, and k the distance in statute miles, at which a line from the height h touches the horizon, taking into account terrestrial refraction, assumed to be of the same value as in the above table, (.075,) we have

$$k = \frac{10}{7.53}\sqrt{h} \qquad h = \frac{10}{17.63}k^2$$

The same result may be arrived at by another method. Instead of finding the correction to be applied to the computed height, find the correction to be applied to the measured altitude. As the measured altitude is necessarily too small, this correction is, under the ordinary state of the atmosphere, additive.

This correction is the angle HAB $= \frac{1}{2}$ ACB. (Book II., Chauvenet's Geom.)

Taking the length of the arc of 1′ as 6086 feet.

Then for any arc k (the distance) the correction

$$\text{HAB is as } \frac{1}{2}\,\frac{k}{6086} =$$

$$0.0001643 \times \tfrac{1}{2}k = 0.0000821k \text{ expressed in minutes;} \qquad (3)$$

expressed in logs., 5.91434 + log. k

$$\text{and the height of BB}' = \tan\,(\text{B}'\text{A H} + \text{cor})\,k \qquad (4)$$

In addition to this correction for curvature, there is also a correction for atmospheric refraction. The rays of light which pass from any distant object on the earth's surface suffer a change of direction, which is called "Terrestrial Refraction," by which the object appears higher than its true position.

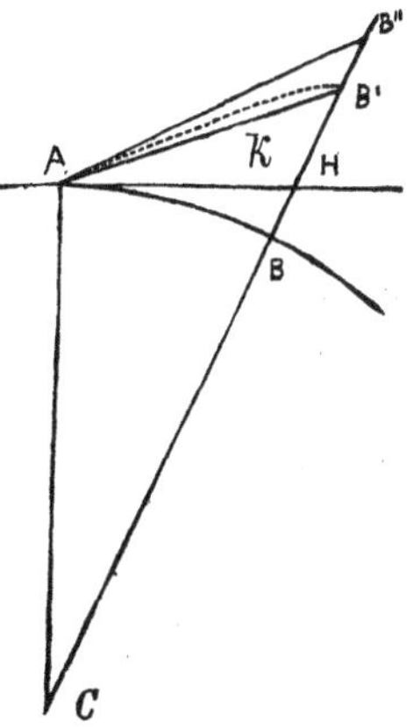

This effect has been determined to be variable, but under the average state of the atmosphere of the sea shore it varies between .074 and ·078. That is, the altitude of the object will appear between ·074 and ·078 (about $\frac{1}{13}$) of the intercepted arc, or distance in *sea* miles, greater than it really is. Thus an object 13 miles distant will be raised 1′ in altitude by terrestrial refraction. The amount of the correction then depends upon the distance of the object, as can be seen from the figure; the object being seen in a line tangent to the curved ray.

The "correction for refraction" may be made in this way. From the measured altitude of object, B″AH, subtract $\frac{1}{13}$ of the distance in sea miles (considered as minutes of arc). Then find from the right angled triangle the height B′H, using the corrected angle B′AH. This will give the height corrected for refraction. To this add the correction for curvature, and the result will give the height of the object above the true level of the observer.

It will be found that the "correction for refraction" by this method will agree very closely with the corrections tabulated together with the "corrections for curvature" on page 54. For instance, if the measured altitude of an object is 2°, distant 5 sea miles, we have

2° log. tan 8.54308
k = 30430 ft. log. 4.48330

Height uncorrected for refraction = 1062.62 log. 3.02668

Subtracting $\frac{1}{13}$ of 5′ = 23″ from 2° we have

1° 59′ 37″ log. tan. 8.54170
k = 30430 ft. log. 4.48330

Height corrected for refraction by method described =
1059.25 log. 3.02500

Correction for Refraction, as taken from Coast Survey formula page 54 . . . . 3.36 ft.

" " by taking difference of the above computed heights , ... 3.37 ft.

The correction for curvature computed by the formula $\frac{k^2}{2R}$=22.165

So that the correction for curvature and refraction = 18.80 ft., and the true height of the object = 1081.42 ft.

The formula $\frac{k^2}{R} \times m$ may be thus deduced. By reference to "Coffin's Navigation," Articles 47, 48, it is shown that the atmospheric refraction decreases the true dip by ·074 of itself. The true dip of the horizon is its distance in sea miles from the observer. From which we learn that the atmospheric refraction increases the true altitude of the horizon or any other object by .074 (or taking the mean co-efficient of refraction .076, which we will hereafter use) of its distance in sea miles.

In the fig. B″B′ is the correction for refraction expressed in feet.

k = AH = AB = AB′, (nearly) =AB″ in feet.

then k expressed in seconds of arc $= \frac{k}{R \sin 1''}$

and the correction of refraction expressed in arc =

$$\frac{k}{\text{R} \sin 1''} \times .076 \qquad (1.)$$

$$\therefore \text{B}'\text{B}'' = \frac{k}{\text{R} \sin 1''} \times .076 \times \text{AB}' \sin 1'' = \frac{k^2}{\text{R}} \times .076 \qquad (2.)$$

To put this in a general form, the correction for refraction (in feet) $= \frac{k^2}{\text{R}} \times m$ in which m is co-efficient of refraction.

46. To find the height of an object, its distance being known, and its altitude being measured above the sea horizon.

From the measured altitude subtract the dip as taken from the dip-tables (corrected for refraction), this will give the apparent altitude above the true horizon of the observer. From this altitude subtract .076 (or $\frac{1}{13}$) of the distance of the object in sea miles. Compute the height of the object by the formula, height $= k$ tan.alt., this will give height uncorrected for curvature, but corrected for terrestrial refraction, compute the correction for curvature, which add to the height already found, and this will give the true height of the object.

As the co-efficient of refraction varies so greatly, it will be seen that observations for the determination of heights by the methods described should only be made when the atmosphere is at its normal condition.

When the water is warmer than the air, the dip is greater than its tabulated value ; when the water is colder than the air, the dip less than its tabulated value.

Prof. Chauvenet (vol. 1, p. 176, Astron.) has deduced the following formula.

$$\text{D}' = \text{D} - 24021'' \frac{t - t_o}{\text{D}}$$

in which t = temperature of the air.

t_o = " " " water.

D being true dip in seconds.

47. To find the distance of an object whose true height and measured altitude above the sea horizon are known.

Estimate the distance. Subtract the dip, as found in the tables, from the measured altitude, and from this subtract $\frac{1}{13}$ of the estimated distance. This will give the altitude above the observer's true horizon, and corrected for terrestrial refraction. Then compute the correction for curvature, which subtract from the true height. With this corrected height and corrected altitude compute the distance from $k =$ cor. height $\times$ cot. cor. altitude.

If k as computed, differs much from the *estimated* distance, it is necessary to approximate again, by using the first computed distance as the estimated distance in the second approximation. By successive approximations a correct result is arrived at.

48. To find the height of an object the altitude of which is measured above a shore horizon.

Prof. Chauvenet, in his Astronomy, vol. 1, art. 126, gives as the expression for the "apparent depression (true dip corrected for terrestrial refraction) of any point of the surface of the water nearer than the visible horizon."

$$\text{Dip} = 22''.14\, d + 39''.07\, \frac{x}{d} \quad (x \text{ being in feet, } d \text{ in statute miles.})$$

or

$$\text{Dip} = 25''.65\, d + 33''.73\, \frac{x}{d} \quad (x \text{ being in feet, } d \text{ in sea miles.})$$

Hence to find the height, subtract from the measured altitude the dip computed by the formulæ just given, as well as the correction for terrestrial refraction due to the distance of the object. Then from this corrected altitude find the height by formula.

$$\text{height} = k \text{ tan. cor. alt.}$$

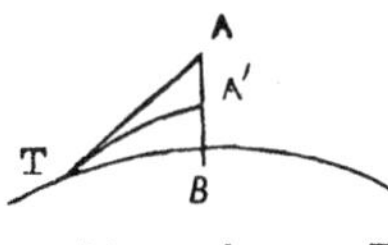

To show how the formulæ given above are deduced, let T be the point the dip of which is required, and A the position of the observer. Let TA′ be a ray of light from T, tangent to the earth's surface at T, meeting the vertical line of the observer at A′.

Put D = the dip of T as seen from A.
d = the distance of T in *statute* miles.
x = the height of observer's eye in feet = AB.
x' = A′ B.

We have from a previous formula deduced in the same treatise

$$x' = \left(\frac{d}{1.317}\right)^2 \qquad (1)$$

and the dip of T as seen from A′ is, according to another previously deduced formula,

$$= 58''.82\sqrt{x'} \qquad (2)$$

both of which are nearly identical with formulæ 53 and 57 Coffin's Navigation.

$$58''.82\sqrt{x'} = 44''.66\, d$$

Now supposing that TA, TA′ to be drawn, the dip of T at A exceeds that at A′ by the angle ATA′ very nearly; and we have nearly

$$\text{ATA}' = \frac{\text{AA}'}{\text{TA}'} \times \frac{1}{\sin 1''} = \frac{x - x'}{5280\, d \sin 1''} \qquad (3)$$

$$\text{whence } \mathrm{D} = 44''.66\, d + \frac{x - x'}{5280\, d \sin 1''}$$

substituting the value x' in terms of d

$$\mathrm{D} = 44''.66\, d + \left(\frac{x}{5280\, d \sin 1''}\right)'' - \left(\frac{d^2}{(1.317)^2 \times 5280\, d \sin''}\right)''$$

since $1'' = .00000485$

$$\mathrm{D} = 44''.66\, d + \frac{x}{d} + 39''.07 - 22''.52\, d$$

$$= 22''.14\, d + 39''.07\, \frac{x}{d} \qquad (4)$$

If d is given in sea miles by exchanging d for $\frac{69\frac{1}{2}}{60}\, d$

$$\mathrm{D} = 25''.65\, d + 33''.73\, \frac{x}{a} \qquad (5)$$

CHAPTER VI.

HYDROGRAPHICAL SURVEYING.

HYDROGRAPHICAL SURVEYING — PLAN OF SOUNDINGS — COAST SURVEY METHODS — DETERMINATION OF POSITIONS — THREE POINT PROBLEM — RUNNING LINES OF SOUNDINGS — PLOTTING SOUNDINGS — REDUCTION OF SOUNDINGS.

49. HYDROGRAPHICAL SURVEYING comprehends the determination and delineation upon a plane surface, of the areas and depths of bodies of water, of the form and character of the bottom, and of the figure and position of the adjacent shores and outlying rocks or shoals, and the representation of natural or artificial marks which serve as aids to navigation. It also includes the observation and reduction of tides and currents, the investigation of which is also styled Physical Hydrography. A Nautical Survey properly comprises both the Topography and Hydrography of a coast or harbor.

PLAN OF SOUNDINGS.

50. In a detailed survey of a harbor, the hydrographic work will be simplified, and made more complete and symmetrical, by the adoption of a plan of sounding lines, prior to the commencement of the work. The selection of the plan falls upon the officer in charge of the survey, and is determined by the degree of accuracy required, the time which may be given to the work, the number of observers, the facilities at hand, and the conditions of winds, weather, and currents. The last named considerations may also influence the determination of the direction of the lines, with a view to the economy of time

and labor. Indeed, all plans will be modified by the peculiarities incident to the survey, and, frequently, a combination of two or more methods will best secure the objects in view. It is not only necessary to designate upon a chart the exact positions and magnitudes of dangers, but also to indicate the undulations and character of the bottom, which frequently enable the navigator, by casts of the lead, to recognize his position, or the course he is making, and which are often at night the only guides to an anchorage. The direction of the lines being determined with reference to the wind, weather, and currents, the number of lines necessary will generally increase with the slope and irregularity of the bottom.

COAST SURVEY METHODS.

51. The following systems for running lines of soundings are used in the hydrographic work of the United States Coast Survey.

System *a* comprises a set of lines normal to a coast, direction of a stream, or line of shoal, at such distances apart as the locality may require. This system furnishes no means by which the positions of the soundings can be verified. Fig. 1, Plate III.

System *b* is composed of a set of lines as in system *a* with another set at right angles to them ; the first giving the longitudinal slope and irregularities, and the second giving cross-sections. Fig. 2, Plate III. This method cuts the field of work into rectangles, and the intersections of the lines provide checks of the accuracy of the soundings. It will generally give all the results required for navigation or nautical purposes.

System *c*. Double lines, running each at an angle of 60° to the line of the coast, or 30° to the general direction of the river, or bay, and making angles of 60° or 120° with each other, with another system, parallel to the coast, or normal to the direction of the river, or bay, crossing the first system of lines at their intersections, so that each of the angles at the point of intersection would be equal to 60° ; all the triangles thus formed by the intersecting lines being equilateral triangles. Fig. 3, Plate III.

This system will give the best results with the fewest lines, and has been successfully used. The soundings are verified by the intersections of three lines.

System *d* contemplates the use of system *b* with another similar system at an angle of 45° ; the two systems having common points of intersection. Fig. 4, Plate III. The space is thus divided into right-angled triangles, the soundings being checked by the intersections of two lines and of four lines. This system is suitable for minute work with large plans.

In making a reconnoissance, or in a rough survey of an unimportant waterway, zigzaging may be resorted to in order to save time ; but, even in this case, the alternate courses should be normal to the stream in order to furnish cross-sections.

The entrance of a harbor should also be sounded by lines run upon the ranges, in order to discover any obstruction in the fair-way.

Dangers, rocks, sunken wrecks, or other abrupt obstructions, and prominent irregularities, should be specially examined, apart from the usual lines of soundings. Alluvial shoals may be developed in connection with the general system.

52. The officers charged with the work of sounding should be fully informed as to the system selected and the degree of detail and accuracy it is designed to secure. Each officer should be furnished with a plot or tracing of the harbor, upon which all the principal stations of the triangulation appear; and it will be found useful to have the plot mounted, and the means at hand of fixing upon it the important points of the lines. It will be convenient to have the watches set to that of the observer of tides. The harbor being duly apportioned among the officers, the soundings will be obtained without loss of time or labor, and without serious omissions.

FIXING THE POSITIONS OF SOUNDINGS.

53. When there is a sufficient number of observers, the observations for fixing the positions of the sounding boat may be made on shore. Two observers are placed at stations of the survey, or at the extremities of any line of the triangulation, and, upon signal being made from the boat, the observers take angles simultaneously upon the boat by sextant, theodolite, or plane table. The times of observation should be noted opposite the angles by watches compared before and after the observations with the watch of the the boat officer. This method is rapid, and gives results which are entirely reliable.

54. Usually the position of the boat is determined by sextant angles taken in the boat upon three points of the triangulation, according to the well known *three point problem*, which is enunciated as follows: *In a given plane triangle, a b c, to find a point, o, such that the three lines*

drawn from this point to the points a, b, and c, shall make given angles with each other. The general solution of the problem is given in Chauvenet's Plane Trigonometry, page 82.

In the application of this problem to the use of hydrographic surveying, the selection of the stations to be observed requires some skill and a knowledge of the principles involved. A consideration of the positions which the observer may occupy with reference to the angles of the triangle will facilitate the practical employment of the problem. The triangle is known in Plane Table practice as the *great* triangle, and the circle whose circumference passes through its angles is called the *great* circle, names which may be conveniently used in this description.

If the sum of the angles x and y, (Fig. 1, Plate IV), observed at the point o, exceeds 180°, the position of the observer will be accurately determined, and evidently lies within the triangle. In this position, when convenient, the third angle z may also be taken, completing *a round of angles,* and furnishing a check upon the accuracy of the observations.

When the sum of the two angles observed is equal to 180°, the observer is upon one of the sides of the triangle, (Fig. 2, Plate IV). Therefore, if the observer, starting from or near any station, as a, or having fixed the position at which the line begins, selects an object as far beyond as possible and directly in range with a second station c, the boat being pulled upon that range, the observer's position at any point on the line may be determined by a single angle from a signal placed at either extremity of the line to a third station b. A single angle x, (Fig. 3, Plate IV), also suffices when the observer is on the prolongation of a side of the triangle. In selecting the third station, it will be sufficient to note that it is as near as possible, and that the angle is not very far from a right angle. In rough water, or whenever the boat has much motion, the angles observed will generally be slightly erroneous; and for any small error in an observed angle, the corresponding error in the position determined therefrom becomes *greater* according as the observed station b is further off, and also as the observed angle becomes smaller.

In Fig. 4, Plate IV, if the sum of the given or observed angles x and y is equal to the supplement of the angle $a\ b\ c$, the problem is indeterminate. In this case, the point o is upon the circumference of the *great* circle. An approach to this condition should be avoided. It cannot occur when the stations observed lie upon the same straight line, or when the middle station observed lies nearest to the observer, the observer being without the triangle, as in Figs. 5 and 6, Plate IV. It is obvious that the stations should be so selected that the observed angles will not be very small.

55. It will, perhaps, assist the inexperienced to consider the practical application of the problem with regard to the common intersection of the circles passing through the position of the observer, and of which the sides of the great triangle are chords. The intersecting circles are shown in the figures already referred to; the geometrical constructions may be found in Jeffers' Nautical Surveying, page 169. The intersection of the circles determines the position of the observer; and the circles themselves, and their consequent intersection, depend upon the values of the observed angles. While, therefore, the angles taken in a small boat, even by careful and experienced surveyors, cannot be considered entirely accurate, the stations should be selected so as to reduce to a minimum the effect of a small error in the observations. In this view, the two circles formed upon those sides of the great triangle, as chords, which subtend the observed angles, should intersect nearly at right angles. In cases approaching the indeterminate, the centres of the circles fall very near each other, and near the centre of the great circle, and the circles are nearly identical. This can only occur when the angle $a\ b\ c$ towards o, Figure 4, Plate IV., is less than 180°, and when the observer is clearly without the triangle. When it appears to the observer that the great circle will pass near his position, one of the stations must be rejected. The positions in which the indeterminate case cannot occur have already been noted; and, generally, it may be avoided by selecting stations so that the intersecting circles will be different either in size or in situation. If the lines joining the middle station with the other stations form nearly a right angle, and the stations are nearly equidistant from the observer, the intersection of the circles will be good; and it will be good in most cases where the centre of each circle is without the other circle. When the middle station is near the observer and the extreme stations relatively far off, the circles will be nearly tangent externally, and their intersections not well determined.

The fixing of points on a line of soundings by one angle, the boat being kept upon a range, requires but one observer, and is accomplished without loss of time. When angles to three points are required, and there is but one observer, it becomes necessary to anchor the boat. When not sounding upon range lines, it is, therefore, better to employ two observers.

RUNNING LINES OF SOUNDINGS.

56. The compass is of little use in sounding, as its indications in a boat are very changeable, even with the improved compasses now supplied to the Navy. Compass lines, run in a tideway, or through the varying currents of a harbor, are also necessarily irregular from these causes. The method by ranges is, however, very simple and reliable, and is especially expeditious in running short lines in a harbor, the soundings being made at equal intervals of time. The boat being pulled at a steady rate upon the range, and the whole distance timed, the work may be made very accurate, and frequent observations for position dispensed with. In harbors where there are not a sufficient number of natural objects, ranges may be put up on shore by the compass, which will ultimately decrease the amount of labor while conducing to the accuracy of the results.

57. The observer fixes the beginning and end of each line, and gives the order "heave" to the leadsman at regular intervals of time, supposed also to correspond with equal intervals of distance traversed by the boat. The frequency of casts, or the length of intervals between them, depends upon the slope and irregularities of the bottom. It is evident that whenever the water shoals or deepens quickly, a small error in the position of the boat will give a large error in the depth of water, when plotted upon the chart. Under these circumstances, it is absolutely necessary that the soundings should be numerous and the positions of the boat continuously determined. With two expert obseivers, the circumstances of this case will not delay the work, but when there is but one observer the boat must be frequently anchored. In anchoring upon a line of soundings, care should be taken to sheer the bcat tideward, or to windward, in order that the boat may ride upon the line.

58. Whatever system and degree of accuracy are employed for the general investigation of a harbor, the approaches to all shoals, detached rocks, and other dangers, must be very carefully determined with reference to the points of triangulation. Having determined the positions of such dangers, and the soundings over them, with extreme care, it is also essential to note the ranges which lead clear of them.

When the existence of isolated rocks has been reported, or is suspected, sweeping should be resorted to. Dangers of this character, steep on all sides, will seldom be revealed by casts of the lead. In trolling for rocks, it is generally best to use two boats with a light line fitted with a chain bight, and to sweep up the slope or steep.

59. In fixing the position and limits of a rock, detached reef, or shoal, a boat may be anchored, or a buoy placed, upon the highest point, and radial lines run to all points of the horizon. Or several buoys may be placed upon a reef to fix the initial points of sounding lines: and if very exact sections of the reef are required, the distance of the soundings from the buoys on the radial lines may be obtained by the use of a marked line one end of which is secured to the buoy used, and the soundings taken with a ballasted pole. When the bottom is muddy, to prevent the pole from sinking into the bottom a disk may be placed upon the end of the pole.

60. Whenever a number of boats are employed in sounding out a harbor, it is important that the lines should be distinguished, and the general direction of each noted, in order that any sounding on the chart may be readily compared with the record. A system of notation for this purpose is given in the instructions for the Deck Board which is supplied to surveying ships of the Navy.

The form for recording soundings, furnished to surveying ships, is given under paragraph 61, together with the instructions for keeping the Boat and Shore book.

61. In work along the coast the soundings may be obtained by running zigzag lines. Observers may be placed upon light-houses or temporary tripods, and the angles are taken simultaneously upon signal being made from the vessel. Theodolites should be used. The angles at the vessel may also be measured for use as checks, and the times should be recorded by watches frequently compared. This is an excellent method where the currents are strong.

FORM OF RECORD OF SOUNDINGS.

U. S. SURVEYING SHIP ....

BOAT AND SHORE BOOK.

DATE : .... PLACE : .... VARIATION : ....

OFFICER IN CHARGE :.... .... LEADSMAN : ... .

NO. OF POS.	TIME.	SOUNDINGS.		COR. FOR TIDE.	ANGLES.			ALTITUDE OF MAST FROM TRUCK TO BAND.		COURSE OF BOAT.	CHARACTER OF SOUNDINGS.	REMARKS.
		Fath.	Feet.	Feet.	° ′ ″	Right Object.	Left Object.	Vessel and Mast.	Angle.			

INSTRUCTIONS FOR KEEPING BOAT AND SHORE BOOK.

1. All the boat and shore work, with the exception of the observations with the zenith telescope, the declinometer, and the dip circle, and of the tidal observations, are to be recorded in the Boat and Shore Book.

2. A book furnished to an officer will be used by him until it is filled. The numbers of the volumes, however, will be current, without regard to the officers using them.

3. The manner of recording will, in general, be the same as that of the *deck-board*, under the instructions given for it, which are to be referred to.

4. The officer, on leaving the ship for boat or shore work, will *set* his watch by the face of the deck time-piece, so that both show the identical time, and on returning he will compare it with the same, and note in the column "Remarks" the time then shown by both. If there is no difference between them he must state this fact.

5. If there be only one observer in the boat, so that simultaneous angles cannot be obtained, the boat must, when on soundings, always be anchored by a grapnel, and must swing to the tide before the angles be taken for position. When not on soundings the boat must be made stationary by holding water with the oars while angling.

6. While angling in sight of the ship or of a shore party, a flag must be exhibited conspicuously, in order that angles may be obtained from the ship or shore party to the boat. Those parties must keep a lookout for such signals.

7. Angles for position must be obtained whenever the boat changes her course, and at every sudden change of the depth of water. The lines between angles for position should never exceed one half-mile.

8. When there is no certainty that the position of the objects angled upon is well determined, and the ship is in sight, the subtension of her mast and an angle between her and one of the shore objects should be obtained in addition to the other angles.

9. The soundings should be at equal intervals at the call "Heave" of the recorder. The time, however, must be recorded at each sounding, and in addition the time of dropping the anchor or checking the boat, and of starting again.

10. Theodolite angles at shore stations will be read by both (if there are four, on the two opposite) verniers of the horizontal circle, and the altitudes of peaks (or their depression, if the theodolite station is on a higher peak), also by both verniers of the vertical circle, the verniers to be distinguished by letters marked on them, the former by A and B, the latter by C and D. They are to be recorded as follows, the headings of the respective columns being changed (in writing) accordingly: Vernier A in the column, ° ′ ″; Vernier B in the column, "Right Object."

The object sighted in the column, "Left Object" (the word "Left" to be stricken out). Vernier C in the column, "Vessel and Mast;" Vernier D in the column, "Angle."

In order to make sure, in obtaining a greater number of angles, of the steady adjustment of the theodolite to the primary object, the latter must be resighted from time to time, and each reading recorded, even if it does not differ from the original. The level must also be re-examined, especially when altitudes are obtained. For the correction of the latter, the elevation of the theodolite station above the level of the sea must be ascertained, and is to be recorded opposite the round of angles, in the column "Remarks," immediately, if ascertained by direct measurement or estimated, or subsequently, as soon as determined by other means.

With a carefully adjusted theodolite the elements for an astronomical bearing may also be obtained correctly enough. The limb observed will be recorded by the letters N. L. or F. L. (nearest or furthest), and L. L. or U. L. (lower or upper).

11. After the return of the party to the ship, the officer in charge will return the book to the navigation officer; the latter will note on each page the correction to be applied for shortness or excess of lead line, and present it to the commander for examination and approval.

PLOTTING SOUNDINGS.

62. When the observations for fixing the positions of the soundings have been taken on shore, the reduced soundings are readily plotted by protracting the angles observed and marking the intersections. It will be necessary to inspect the recorded times to ascertain that the angles were observed simultaneously.

63. If the angles have been taken in the boat, the positions are usually plotted by means of an instrument called the *station pointer;* but the geometrical constructions may be employed to place soundings of great importance. The station pointer consists of a graduated circle to which is attached three arms or radii, the central arm being fixed and its edge coincident with the zero of the graduation. The other arms are movable and have verniers attached by which their edges may be set at desired angles to right and left of the central arm. The lateral arms being thus set to the observed angles, the instrument is placed on the chart, with the edge of the central arm upon the middle station observed, and the edges of the lateral arms bisecting the marks upon the chart which represent the positions of the other two objects observed. The centre of the instrument is then at the point of observation, which is lightly marked upon the chart through the

aperture provided for that purpose. Tracing paper, with a graduated circle printed upon it, or tracing linen on which the angles have been ruled, are good substitutes for the station pointer, and are somewhat more easily used.

64. Whenever the direction of the line does not vary and the soundings change gradually and at a uniform rate, the labor of plotting may be much decreased. It is then only necessary to plot by the station pointer the important soundings of the line and place the soundings between such important positions at equal distances. To avoid the use of an unnecessary number of soundings on the chart, when they are numerous, every second, or third one, may be omitted. Soundings taken at equal intervals are very easily and quickly plotted by the use of a set of equidistant parallel lines, drawn on tracing paper and separated not more than the least distance by the chart scale between any two soundings of a line.

65. Upon the harbor and bay charts of the United States, the contour lines, or lines of equal depths, are traced for every fathom up to three fathoms. This is effected, on the Coast Survey Charts, by variations in the shading, which is made lighter for each fathom as the depth increases to three fathoms, outside of which there is no shading. On the section charts of the coast the contours, generally to the depth of one hundred fathoms, are indicated by regularly dotted lines.

It is a sufficient test of the completeness of the soundings to ascertain if these curves can be drawn on the chart for all depths, without leaving doubt as to their directions anywhere. In the smooth water of harbors, the depth not exceeding 12 fathoms, the difference between soundings at the intersections of lines should not exceed one foot.

Specimens of contour lines are shown on Plate V.

66. The following letters and symbols are employed for the United States Coast Survey Charts to indicate the character of the bottom.

The materials of the bottom are expressed by capital letters. Thus, M stands for mud, S for sand, G for gravel, Sh for shells, P for pebbles, C for clay, St for stones, Gs for grass, &c. Colors or shades are indicated by two small letters, as bk for black, yl for yellow, bu for blue, dk for dark, gy for grey, lt for light, &c. Other qualities of the bottom are denoted by three small letters, hrd for hard, sft for soft, fne for fine, crs for coarse, stk for sticky, brk for broken, sml for small, lrg for large, &c. A combination of these abbreviations states at once the material, color, and other qualities of the bottom, gy. S. brk.

Sh. for grey sand with broken shells, yl. C. St. for yellow clay and stones, fne. dk. S. P. for fine dark sand and pebbles, &c.

Specimens of symbols used on charts to indicate the position of banks, reefs, &c., and some others, are given on Plate V.

In the description of buoys, H. S. stands for horizontal stripes, B for black, and R for red. A buoy with black and red horizontal stripes is a danger buoy. In entering a harbor, black buoys are left on the port hand and red buoys on the starboard hand. A buoy with black and white perpendicular stripes is a channel buoy.

REDUCTION OF SOUNDINGS.

67. As the soundings are taken at all stages of the tide, the times must be referred to that of the tidal observer. The reduction then consists in subtracting from the recorded soundings the number of feet the surface of the water lies above the selected plane of reference, as shown by the tide guage. For this purpose the height of the tide should be observed at stated intervals throughout the progress of the sounding work. It is important to notice that to obtain the proper corrections, reference must be made to a tide guage in the immediate vicinity of the area to be sounded, so that the guage and soundings will be affected by the same form of tide. The character of the tide in the interior of a harbor will generally be different from that of the tide which visits the entrance; and this will be particularly marked where the channel is tortuous, or its natural contractions and expansions influence the drift of the tidal current. The luni-tidal interval and the range of the tide, under these circumstances, will generally be altered in its progress to the second station. A single guage may be used, however, to reduce the soundings made over a large area, throughout which the depth of water is good, and the obstructions to the progress of the tide wave are not numerous or very abrupt. Under these circumstances, the soundings obtained at a distance from the tide guage may be corrected by first determining the difference in the time of tide at the two places, by Airy's rule: *The rate at which the tide travels is equal to the velocity acquired by a body falling through space from a height equal to half the depth of water.* The time of the tide must be reduced to that of the same tide at the guage, and the guage may then be used to reduce the soundings obtained at the distant station. Airy's rule may also be used to determine the distance for which the difference in tide time may be rejected, and the observed soundings referred at once to the tide guage.

68. Where the area to be surveyed is of large extent, it is usual to establish several tidal stations. If the soundings are to be included upon the same chart of the harbor, the soundings should all be reduced to the same plane, by connecting the guages and their benches. This may be done by levels, when the ground is not unfavorable. The following rules for connecting tidal stations are enjoined in the U. S. Coast Survey instructions.

Where there is no considerable river-outflow, make observations upon the two guages to be referred, every fifteen minutes, from one low water to the next following (for the same tides); the average in each case will be *mean level*, and *the difference between the readings of mean level on the two guages is the difference in the elevation of their zeros.*

The above rule will not apply where there is considerable river outflow; but the following rule holds good wherever the outflow of river-water does not wholly reverse the flood-current.

Set up graduated staves at such distances apart that the slacks of the tidal currents shall extend from one to another. By simultaneous observations ascertain the difference in the readings of these guages at the slack between ebb and flood currents, and again the difference at the slack between flood and ebb; *the difference in the elevations of the zeros of the gauges is equal to one-half the sum of the differences of their readings at the two slack waters.* (Appendix No. 11 of the Annual Report of the Coast Survey for 1870.)

It will be seen that this latter rule includes all the cases of the former likely to occur in any single season's work; and since its application involves less time and labor, it may be properly used, whether there is any river slope to be eliminated or not. This same rule will be found applicable to the case where guages inside and outside of inlets are to be connected, and where those above and below obstructions are to be referred to each other.

CHAPTER VII.

TIDAL OBSERVATIONS—CURRENT OBSERVATIONS—REDUCTION OF CURRENT OBSERVATIONS—SAILING DIRECTIONS—TIDE GUAGES—BENCH MARK.

TIDAL OBSERVATIONS.

69. The soundings upon a chart indicate the depths with reference to a selected plane, generally that of mean low water. Tidal observations are, therefore, necessary in connection with a nautical survey, to fix the plane of reference and to obtain the data for reducing the soundings to this plane. The observations consist in measuring by a tide gauge at stated times the heights of the water above a fixed zero of the gauge. The same observations, when continued throughout a sufficient interval of time, furnish the data for the construction of a tide-table, a necessary accompaniment of the completed chart. In order to determine with considerable correctness the elements of the tide, from which the tide-table is computed and the position of the plane of reference calculated, the observations must be carried throughout a lunar month, the local time and the height of water being recorded with especial care at the times of high and low water. The results are but approximations, more or less reliable according to the circumstances of weather during the period of observation. Winds and freshets may temporarily obliterate the usual features of the tide; and in many places long-continued observations are requisite for even a tolerable knowledge of the tidal elements.

For the purpose of reducing soundings, the tide gauge must be noted frequently whenever the soundings are being made.

70. The following is the form of tide-table usually appended to the Coast Survey Charts:

Corrected Establishment at ——————— . . .	$0^{H}XXXII^{M}$
Rise of Highest Tide observed above the plane of reference .	2.6 Ft.
Fall of Lowest Tide do. below do. do. . .	1.8 "
Fall of Mean Low Water of Spring Tides below the plane of reference	0.2 "
Height of Mean Low Water of Neap Tides above the plane of reference	0.4 "
Mean Rise and Fall of Tides	1.4 "
Mean do. of Spring Tides	1.9 "
Mean do. of Neap Tides	0.7 "
Mean Duration of Rise } Reckoning from the middle of one .	5^{h} 59^{m}
Mean do. of Fall } stand to the middle of the next . .	6 19
Mean do. of Stand	0 35

71. The Coast Survey forms for the observation and reduction of tides are annexed.

FORM OF RECORD OF TIDAL OBSERVATIONS.

Observations of Tides at.......................... ....Year............Month..................Day of Month........

Mean Time of Obser.		Reading of Tide Staff.		Wind.		Barometer.		Thermometer.				Atmosphere.	Clouds.				Weather.	Remarks.
															Course of			
Hours.	Min.	Feet.	Dec's.	Direc.	Force.	In's	Dec's.	Att'd.	Air.	Water.	Wet Bulb.		Sky Covered in 8th.	Kind.	Upper.	Lower.		

UNITED STATES COAST SURVEY.

TABLE FOR THE REDUCTION OF TIDES—No. 1.—Showing the Times of High and Low Water, and the Heights of High and Low Tides, together with the Time of the Moon's Passing the Meridian of the place, and the Lunitidal Intervals, at............................ during the month of........ .... 187 .

Hydrographical party of .... Reduced by ....

DATE.	MOON PASSES THE MERIDIAN.		MEAN TIME OF—				LUNITIDAL INTERVAL.				HEIGHT OF—				DURATION OF TIDE.				SLACK WATER.		WIND.		BAROMETER. REDUCED.		THERMOMETER.	WEATHER.	REMARKS.
	Mean Time.		H. Water.		L. Water.		H. Water.		L. Water.		H Water.		L. Water		Flood.		Ebb.		H.	L.							
	H.	M.	H.	M.	H.	M.	H.	M.	H.	M.	F.	Dec.	Ft.	Dec.	H.	M.	.	M.	M.	M.	Dir.	Force.	In.	Dec.	Fahr.		

UNITED STATES COAST SURVEY.

TABLE FOR THE REDUCTION OF TIDES—No 2.—Showing the interval between the time of the Moon's .... transit and the time of High Water; and also the Heights of High Water, at .... from observations made between ...., 18

Hydrographical party of .... Reduced by.. ....

Moon's Transit.		Lunitidal Interval. High Water.		Height of High Water.		No. of Observations.	Moon's Transit.		Lunitidal Interval. High Water.		Height of High Water.		No. of Observations.	Moon's Transit.		Lunitidal Interval. High Water.		Height of High Water.		No. of Observations.
H.	M.	H.	M.	F.	Dec.		H.	M.	H.	M.	F.	Dec.		H.	M.	H.	M.	F.	Dec.	

CURRENT OBSERVATIONS.

72. An accurate knowledge of the velocities and directions of the currents which sweep about the coasts and harbors is of great importance to the navigator; the want of that knowledge is the cause of frequent disasters, involving loss of property and life. The interests of commerce are influenced by the changes in the entrances of harbors, which are produced principally by tidal currents. The navigator is, however, more immediately interested in the former subject, and the determination by observations of the set and drift of tidal currents at all stages of the flood and ebb during the several quarters of the lunar month.

73. Tidal currents are local in extent, being affected both in direction and velocity by the existence of obstructions, either temporary or permanent, and by the configuration of the bottom, in the immediate vicinity. It is, therefore, necessary to establish current stations in proximity to the dangers of a port, and wherever it may be desirable to determine the set and drift, in order to decrease the difficulties of the navigation. The elements of the currents in the approaches to harbors are of the first importance.

74. The observations are very simple, and need not be extended beyond a day in calm weather, the results, when reduced to the mean values, being more reliable than those of a series through a semi-lunation of variable weather. The observations should begin at slack water and may end, after twelve or twenty-four hours, at the commencement of the flood or ebb. The vessel should be anchored in the channel and her position fixed when riding to the flood and to the ebb. The velocity of the current should then be found every half hour while it is strong, and more frequently as it fails, using a log line marked to show the drift in nautical miles and tenths. The set of the current is obtained by sextant, taking the angle from the log chip to a station on shore and as distant as possible. If no well-determined signal is in sight, the direction of the current may be determined by compass, carefully recording the deviation. The magnetic course of the current is wanted. The local mean time of each observation, and particularly at the beginning and end of the slack, the direction and force of the wind, and the depth of water at every slack, should be recorded. The Coast Survey forms for current observations are given herewith.

UNITED STATES COAST SURVEY.

HYDROGRAPHICAL PARTY, under ....

Observation of Currents ....

Station occupied ....

ANGLES OF POSITION, { ....
....

TIME.		ANGLES BETWEEN FLOAT AND			VELOCITY, IN MILES, PER HOUR.		WIND.		REMARKS.
Hours.	Minutes.	Names.	R. or L.	°	Miles.	Dec'ls.	Dir.	Force.	

UNITED STATES COAST SURVEY.

HYDROGRAPHICAL PARTY, under..

Abstract of Observatio s of Currents in...

Station occupied..

TIME FROM NEXT PRECEDING HIGH OR LOW WATER.			DIRECTION OF CURRENT.	At what depth.	WIND.		TIME FROM NEXT PRECEDING HIGH OR LOW WATER.			VELOCITY OF CURRENT.		At what depth.	WIND.	
H. or L.	Hours.	Min.			Dir.	Force.	H. or L.	Hours.	Min.	Miles.	Decl's.		Dir.	Force.

75. In order to reduce the current observations of a single day to the mean value, the tide-gauge must be recorded through not less than a semi-lunation, including the day of current records.

The *current intervals* and the *duration* of a current are estimated from the middle time of slack water, and the interval from the transit of the moon to the time of turning may thus be determined. From extended observations it has been found that the time of maximum velocity corresponds to the middle time of the current. The time of maximum velocity for a single ebb or flood may be found from the observed times and velocities, either by interpolation, or, more closely, by plotting the curve of velocities upon a diagram, using the current intervals throughout the duration of the current and the velocities as co-ordinates, and the *lunar interval* of the observed maximum current then becomes known. The middle time of the rise of the tide is simultaneously observed, and its lunar interval is found; and the lunar interval of the middle time of the tide is computed from a long series of observations by taking the half sum of the mean intervals of high and low tides: the difference between these intervals is a correction to be applied to the lunar interval of the observed maximum current to find the mean lunar interval of maximum velocity.

Using the tidal observations for comparison, the drift for spring tides, or for neaps, may be computed from a short series of current observations, by the following rule, which gives a close approximation to correct results: *the velocities of the currents for corresponding hours of different tides are to each other as the rates of rise or fall at those hours.*

76. The current table, if intended to accompany a chart or sketch of a harbor, should be of the annexed form:

Station.		Current Turns after Moon's Transit.		Maximum Velocity.				Duration.		Slack.		Remarks.
No.	Locality.	Flood to Ebb.	Ebb to Flood.	Flood.	Direction.	Ebb.	Direction.	Flood.	Ebb.	Flood to Ebb.	Ebb to Flood.	

77. The annexed table of tidal currents is taken from one of the Coast Survey Charts:

TIDAL CURRENTS.

No.	LOCALITY.	1st Quarter.		Maximum.		3d Quarter.		Flood or Ebb.
		Set.	Drift.	Set.	Drift.	Set.	Drift.	
1	To the Northward of Cove Point.......	N.3°E. S.9°W.	0.7 0.9	N.18°E. S.20°W.	1.3 1.3	N.18°E. S.20°W	0.5 0.7	Flood. Ebb.
2	Mouth of Patuxent River.............	S.45°W. N.70°E.	0.1 0.3	S.67°W. N.59°E.	0.4 0.7	S.75°W. N.62°E.	0.2 0.4	Flood. Ebb.
3	Off Cedar Pt.......................	N.16°W. S.30°E.	0.3 0.2	N.5°W. S.37°E.	0.7 0.4	N.9°W S.42°E.	0.4 0.3	Flood. Ebb.
4	Off Point No Point	N.10°W. S.11°W.	0.3 0.3	N15 W. S.45°E.	0.5 0,4	N.26°W. S.45°E.	0.3 0.2	Flood. Ebb.
5	Hooper's Straits....	N.34°E. S.56°W.	0.3 0.3	N.45°E. S.45°W.	0.5 0.6	N.45°E. S.21°W.	0.3 0.3	Flood. Ebb.

The Observations were made as far as practicable when the influence of the wind was small. The Bearings are true; the Drifts are in nautical miles per hour, shown on the Chart by figures near the ends of the current arrows.

——— —+—> Signifies end of 1st quarter,
——— —++—> Maximum,
——— —+++—> 3rd Quarter Flood.
<—+— ——— signifies end of 1st quarter,
<—++— ——— Maximum,
<—+++— ——— 3rd Quarter Ebb.

The changes in the set and drift of tidal currents under the influence of strong winds, and especially of those winds which prevail at certain seasons, should be observed; and when the results are important, full descriptions of the changes should be entered, as remarks, upon the chart.

SAILING DIRECTIONS.

78. In connection with the views now usually placed upon charts, sailing directions are of great use to the navigator, particularly in approaching unfamiliar coasts. Such remarks upon the chart are intended to bring prominently to notice the dangers of the coast, or harbor entrance, and the means of avoiding them. The courses which lead to the anchorages are given, generally from the magnetic meridian, with bearings of the prominent points and ranges, and descriptions of day marks and night ranges. Subsequent to the survey of a harbor these remarks should be carefully prepared, both for entering and leaving the harbor. No specific rules can be laid down here as guides in this work; and the student is referred for information as to the particulars of the sailing directions to the Coast Survey Charts.

TIDE GAUGES.

79. For harbors and localities which are sheltered from the swell of the sea, the most reliable gauge is a staff graduated upwards in feet and tenths, which is secured in a vertical position with its zero below the level of the lowest tide. The zero of the staff is connected with the bench-mark, or permanent mark of reference, by levellings. If a levelling instrument is not at hand, the reference may be effected by measuring the difference of height of a number of intermediate points by means of a long straight-edged board, made horizontal by the aid of a carpenter's spirit-level, or a plummet square. The differences of height should be determined in both positions of the level. A line of sight from the bench-mark across the tide-staff to the sea-horizon, is an excellent level line.

80. Where the surface of the water is disturbed by ripples, or when nice observations are required, a glass tube of about a half-inch in diameter with a float inside is secured to the face of the staff. The lower end of the tube should be partially closed by a cork with a hole through it, and the tube should extend at all times several feet below the surface of the water, so that the undulations will be excluded.

The empty bulb of a thermometer with a portion of the stem attached to give it weight, makes a good float. A red glass bubble, or a little colored oil, may be used.

81. In places were the swell of the sea is considerable, or when it is inconvenient at times for the observer to see the figures on the staff by reason of the great range of the tide, the *box gauge* is employed. It consists of a box, closed at the bottom, containing a graduated staff or rod attached to a tin or copper float which is moved up and down by the tide. The water is admitted through gimlet-holes on the sides and near the bottom. The staff of a box gauge is graduated downward, and the observer notes the figures on the staff as they pass the top, or a convenient opening in the side of the box, technically called the *reading point*. Besides the reference of the reading point to the bench mark by a line of levels, it is necessary to record the distance from the zero of the staff, or some other stated division, to the water line of the float, in order that the elevation of the bench above the surface of the sea may be computed for the readings of the gauge; and as the water line of the float may change with the saturation of the staff, or other cause of alteration in weight, this distance should be occasionally measured and recorded.

If the holes at the foot of the gauge are too small, or become partially closed by weeds, the ingress and egress of the water will not be free, and the observations become erroneous. On the other hand, if the holes are large, the undulations of the sea will affect the float and staff. The effect of inadequate water holes is to increase the *intervals*, reduce the *ranges*, and alter the magnitudes of all the *inequ lities* of the recorded tide.

82. To the box gauge is sometimes attached a self-registering apparatus, so arranged that the rising or falling float carries a pencil with which it describes a curve upon paper wrapped about a cylinder revolving by clock work. The Saxton self-registering tide gauge is described in the Coast Survey Report of 1853. In this gauge the motion of the pencil is horizontal, and the paper unrolls from one cylinder and rolls up on another, the receiving cylinder being revolved twice in a day, and a mark being pricked upon the paper at each half hour. The curves on the paper are reduced to figures by the use of scales; and in order to determine the absolute heights represented by the curves it is necessary to refer to a staff gauge in the immediate vicinity. The staff observations should be frequently made on quiet days, at even hours, and about the time of *stand.*

The recording pencil should be drawn back and a vertical mark made on the paper, over which should be written the true time and the height by staff. These observations connect the curves on the paper with the bench mark, and furnish a check upon the clock and gauge.

83. Hunt's *Off-shore Tide and Sounding Meter* is an instrument by which the heights of water are determined, barometrically, by the pressure of the sea upon an elastic air bag lying upon the bottom. This instrument is described in the Coast Survey Report for 1857.

BENCH-MARK.

84. The reduced soundings upon a chart detail the depths of water above some selected plane of reference, which is that of *mean low water* upon the Coast Survey Charts. The observations of high and low waters throughout a lunar month determine the vertical distance of the plane of reference from the zero of the tide gauge employed. But as the gauge is liable to injury and decay, and in view of the probability of a subsequent survey of the same locality, it is further necessary to fix the level of the plane of reference with regard to the position of a permanent *bench.* This is effected by running a line of levels from the zero of the tide gauge to the *bench-mark.* Upon a re-survey, the plane of reference of the soundings may be recovered from the position of the bench-mark, and a comparison of the surveys established. This is frequently of great importance in tracing the history of a harbor, particularly in the formation and disappearance of shoals, with a view to the improvement of the tidal basin.

Any object near the shore which is of a permanent character, or is not likely to be removed, may be used as a bench.

The bench should be marked by a circle with cross lines at the centre to indicate the exact position of the reference point. The location of the bench should be carefully described in the tidal record, and a note made of its elevation above the plane of reference.

CHAPTER VIII.

TIDES—DEFINITIONS—ORIGIN AND VELOCITY OF TIDE WAVES—FORMS OF TIDES.

85. Tides.—The surface of the ocean rises and falls twice in a lunar day, about 24 h. 50 m. of mean solar time. The explanation of this phenomenon depends on the law of universal gravitation. The tides are due to the *difference* of the attractions of the moon upon the earth and on the waters of the ocean, and to a similar inequality in the attractions exerted by the sun. To an observer on the coast, the tides appear as alternate depressions and elevations of the sea, and also as horizontal movements of the water, alternately ebbing and flowing; and the term *tide* is ordinarily employed to designate both phases of the phenomenon.

The word *tide* signifies, in hydrography, only the vertical movement of the sea, and the words *rise and fall* are used with reference only to the same movement. The horizontal motion of the water is called the *tidal current*, and the words *flood* and *ebb* indicate the general direction of the tidal current. The word *set* applies to the compass direction and *drift* to the rate or velocity of the tidal current. *Stand* is that interval of time, at high or low water, during which no vertical motion of the water can be observed. *Slack* designates the interval of time during which no horizontal motion is perceptible. *Range* is the difference in the height of the tide from high water to low water; and the term is equivalent to the expression *mean rise and fall.*

The *length* of the tide is the distance measured on the surface of the ocean from one low water to the next.

86. The tide wave must be regarded as the result of the combined action of the sun and moon, though each of these bodies may be supposed to raise an undulation, or wave. But since the influence of the moon in creating tides is known to be $2\frac{1}{3}$ times that of the sun, the tide may be considered as the lunar wave, modified or exaggerated by the influence of the sun, according to the relative position of the two bodies in longitude.

87. Soon after the sun and moon are either in conjunction or in opposition, the highest tides of the lunar month will occur. These are called *Spring Tides*. Unusually high tides will occur when the earth and moon are in the position of their respective orbits, at the time of new or full moon, in which the attractions of the sun and moon are greatest upon the waters of the earth. This may occur when the moon is in perigee, and new or full, about the 1st of January. See Paragraph 96.

88. A third astronomical condition affecting the height of the tide will presently be stated; depending upon the declinations of the bodies and the latitude of the place.

89. *Neap Tides* are the smallest of the lunar month, occurring shortly after the moon is in quadrature. At that period the attractions of the sun and moon upon the waters act in opposition and the solar wave tends to counteract the lunar. The range of the tide, therefore, gradually decreases as the moon passes from conjunction or opposition to quadrature, and increases upon leaving the positions in quadrature. See Paragraph 96. The ratio of the spring tide to the neap tide is about that of 5 to 2.

Very small tides will take place about the time of the earth's perihelion passage if the moon is in apogee and also in quadrature; the influence of the sun in creating tides being then at its greatest, while the moon has its least possible effect.

90. During the first and third quarters of the lunar month, the solar wave lies to the west of the lunar one, and the combined undulation, or tide wave, will be to the west of that which would be due to the moon alone; and this causes a *priming*, or acceleration, of the time of high water.

In the second and fourth quarters the solar influence acts in retardation of the lunar wave, and a *lagging* of the tides takes place.

91. The interval of time which elapses from the instant of the moon's transit over the meridian of a place to that of high water at the same

meridian is called the *luni-tidal interval.* The variations in this interval, from day to day, are due to priming and lagging.

92. There is a daily retardation in the time of high water, following the retardation of the moon. The two daily tides at any place are not, however, always equal in height, owing to the inclination of the plane of the moon's orbit to the plane of the earth's equator and to the diurnal rotation of the earth. When the declination of the moon is zero, the two daily tides at any place will be equal in height, and the tide wave is highest at the equator and lowest at the poles. When the moon is not in the plane of the equator, the highest point of the tide wave, and the greatest daily inequality, will occur at places whose latitude, either north or south, is equal to the moon's declination; and the two daily tides at any place, except upon the equator or at the poles, are unequal. If the declination of the moon is of the same name as the latitude of the place, the greater of the daily tides occurs next after the upper transit of the moon; but if the latitude and declination have contrary names, the higher tide of the day follows the lower culmination of the moon. The position of the sun in declination and the rotation of the earth being considered, results may be stated which are entirely similar to those described above. The difference in the daily tides at a place, due to these causes, is known as the *diurnal inequality.*

93. The mean of the values of the luni-tidal intervals at any place on the days of full and new moon is called the *common establishment* of a port. The *corrected establishment* is the mean of all the luni-tidal intervals in a lunar month.

94. The time of the moon's transit advances from 0 h on the day of new moon to 12 h at the full. And if we observe the changes in the length of the luni-tidal intervals from conjunction to opposition, or *vice versa,* and the changes in the range of the tide during the same period, it will appear that the variations depend closely upon the hours of the moon's transit, the same form of tide wave occurring at equal intervals from the times of conjunction and opposition. This alteration in the figure of the tide during the semi-lunar month is known as the *half-monthly inequality.*

95. Were the whole earth covered with water of uniform depth, the tide wave, which results from the attractions of the sun and moon, would accompany the moon in its apparent diurnal motion, sometimes preceding and sometimes following by small intervals the passage of the moon over any meridian, the mean angular velocities of the wave

and moon being equal. The tide wave, under such circumstances, would be a simple undulation, without progressive motion of the waters. The irregularities of the bed of the ocean, the interposition of continents, and the action of winds, are primary causes of the great variations in the velocity and direction of the tide wave and of the manifold forms of tidal currents. The actual phenomena observed are extremely complicated, differing widely from theoretical conclusions; the explanation of which, and of their laws of dependence, requires long periods of careful observation and a discussion of the physical sciences. It is necessary only to state here some of the results obtained.

96. An examination of the half-monthly inequalities, drawn from tide records, discloses the fact that the mean luni-tidal interval and the highest tides do not take place till a day or two after the times of full and change; and, further, this *retard* of the tide changes with the locality. It is evident, therefore, that the passage of the moon over a meridian is not generally the cause of the succeeding high water. The *retard, or age of the tide*, at any place, is the interval of time from the formation of the tide wave in mid-ocean till its arrival, or culmination, at that place.

97. If lines are drawn upon a map connecting all those places at which high water occurs at the same instant of absolute time, such curves are known as *cotidal lines*. Usually upon these maps, numerals are placed over the cotidal lines, which indicate the hours of Greenwich time on the days of full and change at which high water occurs on the different lines. An inspection of such a map will show that the great ocean tide wave originates in the Pacific, west of the South American coast. Thence, travelling to the northwest, it reaches the coast of Kamtschatka in about ten hours. The same wave, travelling with less velocity through the more shallow water of the South Pacific, passes New Zealand in about twelve hours, traverses the Indian Ocean to the Cape of Good Hope, turns into the Atlantic, and finally reaches the coast of the United States in about forty hours from its formation. Another branch of the tide wave is propagated towards Cape Horn, and turns eastward into the Atlantic.

98. The velocity of the tide wave appears to be determined mainly by the depth of the water through which it is propagated. The range depends more upon the variations in width than in depth, and upon the peculiarities in the configuration of the coast. Referring to Airy's rule, already stated in Chapter VI, Paragraph 67, the velocity of the tide wave for different depths of the ocean is as follows:

Depth of Water.	Velocity of Wave per hour.
25 feet.	19 miles.
100 "	39 "
250 "	61 "
1000 "	122 "
5000 "	273 "
20000 "	547 "
50000 "	865 "

99. The form of tide at any place may be graphically represented upon a diagram, by plotting with the hours of the day as abscisses and the heights of the tide as ordinates ; and the tides of different localities may be thus classified. Three principal forms of tides are now recognized, with reference to their figure and causes of origin.

Along the eastern coast of the United States the *semi-diurnal* wave prevails, giving two high and two low waters in a day. It is supposed that the great tide wave from the South Atlantic arrives on our coast twelve hours after the arrival of the direct wave which crosses the Atlantic from east to west, the higher tide of the former coinciding with the lower high water of the latter, thus eliminating the diurnal inequality.

In the Gulf of Mexico the semi-diurnal tides are very small, being nearly destroyed by the interference of the tide wave which passes through the Straits of Florida with that which reaches the Gulf through the Yucatan Channel. The tides are there of the *diurnal* form, one of the daily tides being nearly inappreciable, and there being but one high and one low water in a day. Tides of the diurnal type have but a small range, which nearly disappears when the moon is on the equator.

The *compound* form of tide is observed on the Pacific coast of the United States, where two daily tides usually occur, which are very unequal, except when the declination of the moon is zero. At Port Townsend, near Vancouver's Island, when the moon's declination is greatest, this inequality reaches a maximum, one of the daily tides being obliterated. There is, at such times, no descent corresponding to one ebb, but only a check in the rise of the tide.

100. On the east coast of Scotland there is an appearance of high water four times a day, due to the superposition of two tidal waves, one passing through the English Channel,and the other travelling around the north of Scotland and turning southward into the North Sea. Many other tidal phenomena have been observed, but it is designed here to refer only to the more simple ones. The tides of rivers are exceedingly complex, and are not well understood. The whole subject is one for patient and extended investigation, for which the data must be obtained by careful and long continued tidal observations.

CHAPTER IX.

PRACTICAL HINTS.

HINTS FOR THOSE WITHOUT EXPERIENCE IN THE PRACTICE OF HYDROGRAPHIC SURVEYS.

101. Care in the preparation and preservation of data. Without the note books are sufficiently full, and capable of being plotted in detail by any intelligent surveyor, the work is of NO VALUE. The topography made is the only thing that can be excepted.

102. However well a surveyor may understand the theory, the practice or execution is ever varying. The *purpose* of every survey must determine its minuteness, and in general its method of execution. If it is intended as a preliminary to some improvement, the greatest exactness, and a large scale for plotting is requisite; if only for safe navigation, the character of the bottom and the dangers to be guarded against must be sufficiently shown by ranges, if possible, or by defining the position for necessary buoys by soundings, or by the character of the bottom, and all of these elements will usually enter into the survey of an ordinary harbor.

103. Should a survey for improvements be required, the careful measurement of a *base line* in its immediate vicinity is essential; and if possible, in such a manner as to mark distances apart upon which *range lines* are to be run, and a parallelogram, formed with the range points established, marked by stakes numbered or lettered, to ensure persons bearing *portable ranges* to take up the designated positions when the soundings are made; the observer in the boat taking as many horizontal angles when sounding, as is found necessary to fix in position properly, the depth and the character of the bottom. Some-

times it will be found advantageous, in advance of beginning to sound, to have the men bearing the portable ranges to take up their positions consecutively, on the proposed lines, and to place on the shore line small flags of different colors to be taken up as the work progresses to avoid confusion.

104. A special study of every bar or harbor in advance is necessary, based upon the best chart or information known, the object being to define most, and with accuracy, with the fewest soundings and the least work; range lines on trigonometrical or other established points should be run, and for this purpose points should be established; these range lines should have a horizontal angle *taken over the lead*, at every other cast and should be plotted in ink of a different color from common soundings.

105. Well-defined natural ranges on trigonometrical points should be run in such directions as to best fill in the work. If the range is kept closely on, and the ends of the line determined by two horizontal angles, or by one, and a cut upon the boat from a shore station, this work may be regarded as of almost the same value as the former.

106. To guard against the effects of winds and tides all lines should be run, if possible, on natural ranges, even when neither point of the range is determined trigonometrically. In "filling in" by the three point problem it is necessary to mark closely where the work must come in and to take a tracing of the chart as a guide, if it is to be done in a boat. The traverse made is usually from 15 to 20 degrees, and will be made more satisfactory by laying off *with a sextant* the necessary angle just before changing the line; by varying this angle slightly a good natural range can usually be secured, whether running off or on shore, the first on a back range, the second direct.

107. In the completion of the work, if possible, range lines should be run on prominent natural objects, so as to define, as nearly as can be done, the limits of the channels, and these lines should form the bases of sailing directions, which should be written at the time of completing the survey, and the vessel run by them and her track plotted.

108. The *character* as well as the depth should be called at every sounding; an examination of charts in general will show that soft bottom will be found in mid-channel; that upon one side it will shoal gradually in mud, and upon the other side *hard bottom* will usually indicate the channel limit, *before shoaling* even, and that it shoals rapidly to a shore, spit, or bank.

109. A very useful movable base may be established on board every

vessel to determine distances within two or three miles with considerable accuracy by carefully measuring the longest line of sight obtainable on deck, and putting in a clump of copper tacks to mark the ends distinctly and permanently, and calculating a table in advance, corresponding to the base in length and the forward angle. The after angle will be made 90 degrees by means of the helm, between the object and the observer forward; his angle being between the object determined upon and the observer aft, taken at the precise time indicated by the observer aft.

110. No survey showing an approach from the sea to the land that has not the neighboring highlands and peaks located and the height approximately marked on them, is complete. An approximate height is so readily determined that there is no excuse for its omission.

CHAPTER X.

PROJECTION BY DEVELOPMENT.

PROJECTIONS — DEFINITIONS — PROJECTION BY DEVELOPMENT—FLAMSTEED'S AND BONNE'S PROJECTIONS—ORDINARY POLYCONIC OF U. S. COAST SURVEY—EQUATIONS OF A MERIDIAN—COMPARISON OF THE PROJECTIONS OF CURVES WITH THE CURVES ON SPHEROID—SAME AS TO ANGLES —PROJECTION OF THE SHORTEST DISTANCE BETWEEN TWO POINTS.

111. Suppose a polygon to be inscribed in the elliptical meridian so that each side subtends an equal meridional arc. By revolution about the polar axis the ellipse will describe the spheroidal surface and the polygon will describe a series of frustums of cones. Evidently the two surfaces will equal each other when the arc of the curve equals its chord. In other words, we can substitute for the elementary spheroidal zone, the elementary conical surface. From this point of view the spheroidal surface is formed by the intersection of an infinite number of cones tangent to the surface, along successive parallels of latitude. These conical elements may be developed on a plane, and according to the law of their development, shall we have peculiar properties in the resulting chart. If the elements are developed parallel, and adjacent to each other, it follows that the resulting surface will be *equal* to the surface of the spheroid. As the elements may, under this condition, be either straight or curved; so we may have any number of charts of a given spheroidal surface equal in area, and differing in the curve in which the parallel and adjacent elements are developed.

FLAMSTEED'S PROJECTION.

112. As instances of two charts where the spheroidal area is represented by an equal chart area we instance Flamsteed's and Bonne's projections. See fig. 1, Plate VI for Flamsteed's.

Here a straight line, N S, is taken equal to the rectified arc of the meridian, between the latitude, L and L′ (in this case equals $\frac{\pi}{2}$) This line represents the principal or middle meridian; as the conical elements are developed equally to the right and left of N S, and are perpendicular to it.

The conical elements here become rectangular elements; the circles of latitude being developed as parallel straight lines, perpendicular to the middle meridian.

The parallels are developed with their proper lengths; therefore, to trace the meridians differing K n° (K being any number), in longitude from the middle meridian; divide the parallel whose length on the sphere $= 2\,\pi \times a \cos l$ (a being radius) into $\frac{360}{n^\circ}$ parts, one of these distances, measured in its proper latitude on chart perpendicular to N S, will give the intersection of the first meridian with such parallel. Any multiple of this distance will, in like manner, give the position of a point on the meridian K n° from the middle meridian; by laying off points on the different parallels we may trace the meridians through them. These meridians will be curves. Evidently if we take S for the origin, S N and S E axes of x and y, respectively, we shall have for the position of any point, whose longitude from N S is n° and whose difference of latitude from S is l

$$x = a\,l\,\frac{\pi}{180}$$

$$y = \pi\,a \times \frac{n^\circ}{180^\circ}\cos l$$

which are the equations of a meridian when n is constant.

BONNE'S PROJECTION.

113. In Bonne's projection (see fig. 2, Plate VI) the conical elements are disposed of in concentric arcs of circle. The common centre of these arcs lies in the middle meridian; produced towards the pole

and at a distance from the parallel of the middle latitude equal to $a \times$ cot. latitude. NS, as before, is the rectified arc of the meridian, between L & L′, in this case 0° and 90°. The parallel of 45° is described with radius a cot. 45°, and the other parallels are described from the same centre with radii equal to a cot. 45°$\pm$ rectified arc of meridian between latitude 45° and that of the required parallel. The other meridians are traced by obtaining points in the same way as for Flamsteed's projection.

Flamsteed's projection is only a particular case of Bonne's, where the middle latitude is 0°.

In Bonne's projection the meridians will be curves. The middle meridian and all the parellels cut each other at right angles and are projected with their true lengths. The *middle parallel* cuts all the meridians at right angles. The other parallels and meridians do not cut at right angles. In short, the projection is orthogonal along the middle meridian and parallel.

Bonne's projection is best adapted to represent a narrow zone, as the meridians will cut the middle parallel at right angles.

Where there is considerable difference in latitude and a moderate difference only of longitude to be projected, it is best to make use of the ordinary polyconic, as the commendable characteristics of this projection are independent of change in latitude, in which respect it differs from Bonne's projection.

ORDINARY POLYCONIC.

114. In the ordinary polyconic (see fig. 3, Plate VI.), as used by the U. S. Coast Survey, a middle meridian is taken as for the preceding cases. The conical elements are disposed in arcs of circles described with radii equal to $a \times$ cotangent of the latitude. The centres of these arcs lie in the middle meridian produced, each arc cutting the middle meridian at its proper latitude.

It will be seen that these elements touch each other only at the middle meridian, since they are described from different centres, they therefore diverge from each other as they leave it, and consequently equality of areas is not preserved. The parallels are projected in their true lengths as before and the meridians are traced in the same way.

ELLIPTICITY OF EARTH.

115. To allow for the ellipticity of the earth (see fig. 4, Plate VI)

we must use for the radius of the developed parallel N cot l, where

$$N = \frac{a}{(1 - e^2 \sin^2 l)^{\frac{1}{2}}},$$

or is the normal terminating in the minor axis and l is the angle it makes with the major axis. The rectified middle meridian will also be different in this case, as equal meridional distances do not correspond to equal differences of latitude on the ellipse. N cot $l = r$ is evidently the slant height of the tangent cone.

LENGTH OF DEVELOPED PARALLEL.

The parallels on the spheroid have radii equal to N cos l and the length of any arc $n°$ of a parallel will be $n° \frac{\pi}{180°}$ N cos l.

SCALE OF PROJECTION.

116. In making projections it is, of course, impossible to make charts equal in area to the surfaces they represent. They must be many times smaller for convenience of construction and reference. In practice, they seldom equal $\frac{1}{5000}$ the real size, and then only when the area to be represented is only a few square miles in extent.

This reduction is effected by multiplying the elements of the spheroid by the co-efficient of the scale as $\frac{1}{1500000}$. We shall thus reduce all magnitudes to manageable dimensions and we shall really produce a chart of a spheroidal surface; similar in every respect to that of the earth, but only $\frac{1}{1500000}$ as large.

CONSTRUCTION OF PARALLELS AND MERIDIANS.

117. In practice, it will be found inconvenient to describe the arcs of the parallels with radii.

They can better be drawn by constructing them from their equations which are those of circles. We can thus not only construct the parallels, but can determine the points in which the meridians intersect them and then trace the meridians.

It will be found more convenient to have x and y, the rectangular co-ordinates of a point, expressed as functions of the radius of the parallel and the angle the radius makes with the middle meridian.

Plate VI, fig. 5. Let θ be this angle. Then taking the origin at L, at the point of intersection of any parallel with the middle meridian, taking the middle meridian as axis of y and the perpendicular through L, as the axis of x, we shall have for any point P, whose latitude is L, and longtitude from the middle meridian n°.

$$x = \text{Y P} \sin \theta = \text{N} \cot l \sin \theta \quad (1)$$

$$y = \text{Y P versin}\, \theta = \text{N} \cot l \text{ versin}\, \theta \quad (2)$$

where θ is some function of n evidently.

To determine the relation of n and θ, it is only necessary to remember that the parallels are projected with their true lengths; in other words, the *distance* LP, equals the distance between the points L and P on the spheroid, measured on the arc of the parallel passing through L and P.

Therefore, the angles at the centres of the two arcs will be in inverse proportion to the radii or

$$\text{N} \cot l \times \theta = \text{N} \cos l \times n^\circ$$

$$\therefore \qquad \theta = n^\circ \sin l \quad (3)$$

These three equations are sufficient to project any point of the spheroid, given by its latitude and longitude from the middle meridian.

By taking n constant, we can find successive points in the projection of any meridian.

GENERAL EQUATIONS OF A MERIDIAN.

118. Taking the origin at L, the general equations of a meridian will be

$$x = \text{N} \cot l \sin \theta \quad (4)$$

$$y = \pm \text{S} + \text{N} \cot l \text{ versin}\, \theta \quad (5)$$

$$\theta = n \sin l \quad (6)$$

where + S is the distance, towards the pole, on the middle

meridian from the origin to the point where the parallel, of the point to be projected, cuts the middle meridian. It is therefore equal to the arc of the elliptical meridian between the latitudes of origin and point.

ANGLES MADE BY MERIDIANS WITH MIDDLE MERIDIAN.

119. Let φ be the angle the meridian makes with the middle meridian

$$\text{then } \tan\varphi = \frac{dx}{dy}$$

$$dx = \cot l \sin\theta \, d\,\mathrm{N} + \mathrm{N} \cot l \cos\theta \, d\,\theta - \mathrm{N} \sin\theta \operatorname{cosec}^2 l \, dl$$

$$dy = ds + \mathrm{N} \cot l \sin\theta \, d\,\theta - \mathrm{N} \operatorname{versin}\theta \operatorname{cosec}^2 l \, dl + \cot l \operatorname{versin}\theta \, d\,\mathrm{N}$$

$$d\,\mathrm{N} = d\,\frac{a}{(1 - e^2 \sin^2 l)^{½}} = \frac{ae^2(1 - e^2 \sin^2 l)^{-½} \sin l \cos l \, dl}{1 - e^2 \sin^2 l}$$

$$= \frac{ae^2 \sin l \cos l \, dl}{(1 - e^2 \sin^2 l)^{\frac{3}{2}}} = \frac{\mathrm{N}^3}{a^2} e^2 \sin l \cos l \, dl$$

$$d\vartheta = \dot{n} \cos l \, dl \qquad ds = \mathrm{R}_m \, dl \qquad \text{where}$$

$$\mathrm{R}_m = \text{radius curvature meridian} = \frac{a\,(1 - e^2)}{(1 - e^2 \sin^2 l)^{\frac{3}{2}}}$$

see calculus, radius of curvature.

$$\therefore \tan\varphi =$$

$$\frac{\frac{\mathrm{N}^3}{a^2} e^2 \sin\theta \cos^2 l + \mathrm{N}\,n \cot l \cos\theta \cos l - \mathrm{N} \sin\theta \operatorname{cosec}^2 l}{\mathrm{R}_m + \mathrm{N}\,n \cot l \sin\theta \cos l - \mathrm{N} \operatorname{versin}\theta \operatorname{cosec}^2 l + \frac{\mathrm{N}^3}{a^2} e^2 \operatorname{versin}\theta \cos^2 l}$$

$$= \frac{\frac{\mathrm{N}^2}{a^2} e^2 \sin\theta \cos^2 l + n \cot l \cos\theta \cos l - \sin\theta \operatorname{cosec}^2 l}{\frac{\mathrm{R}_m}{\mathrm{N}} + n \cot l \sin\theta \cos l - \operatorname{cosec}^2 l \operatorname{versin}\theta + \frac{\mathrm{N}^2}{a^2} e^2 \operatorname{versin}\theta \cos^2 l} \qquad (7)$$

$$\text{If } e = 0,\ \mathrm{N} = \mathrm{R}_m = a$$

$$\therefore \tan \varphi = \frac{\theta \cos^2 l \cos \theta - \sin \theta}{\theta \cos^2 l \sin \theta - \text{versin}\, \theta + \sin^2 l} \qquad (8)$$

gives the angle the meridian makes with the middle meridian on the polyconic projection of the sphere. Taking the case where $l =$ 45° and $\theta = 3°$ which gives $n = 4.°2426$

For the spheroid $\varphi = 177°00'04.''9 +$
For the sphere $\varphi = 177°00'04.''9 +$

which values agreeing to the tenth of a second may be considered equal.

$$\text{Since } \theta - (180° - \varphi) = 4''.9$$

the projection may be considered orthogonal.

ANGLES BETWEEN MERIDIANS AND RADIUS OF PARALLEL.

120. Let $\mu = \theta - (180° - \varphi)$, then $\tan \mu = \dfrac{\tan \theta + \tan \varphi}{1 - \tan \theta \tan \varphi}$

$$\therefore \tan \mu = \frac{\theta - \sin \theta}{\sec^2 l - \cos \theta} \qquad (9)$$

which will give correct values for the projection of the sphere, for all values of l and n. If n be not greater than 3°. (9) will be sufficiently exact for the spheroid also.

Evidently μ will be zero at the pole.

When $l = 0$ the expression takes an indeterminate form $\frac{0}{0}$ but substituting for $\sin \theta$ and $\cos \theta$

$$\tan \mu = \frac{\theta - \theta + \frac{\Theta^3}{6} +}{\text{Sec}^2 l - 1 + \frac{\theta^2}{2} +} = \frac{n^3 \sin^3 l +}{6\left(\tan^2 l + \frac{n^2 \sin^2 l}{2}\right)}$$

$$= \frac{n^3 \sin l}{6 \sec^2 l + 3\, n^2} = 0 \text{ when } l = 0$$

For latitudes 34°, 35° and 36°, μ is less than one second of arc when n equals three degrees.

ELONGATION OF MERIDIANS.

121. For the sphere

$$dy = dl\,(1 + n \cot l \sin \theta \cos l - \text{versin}\,\theta \operatorname{cosec}^2 l)$$

and as θ and φ are sensibly equal within three degrees of the middle meridian, since

$dy \sec \varphi = d\,m$, m being the projected meridian

$$d\,m = dy \sec \theta = dl\,(\sec \theta + n \cot l \tan \theta \cos l - (\sec \theta - 1) \operatorname{cosec}^2 l)$$

$$= dl\,(\sec \theta + \cot^2 l \tan \theta.\ \theta - (\sec \theta - 1) \operatorname{cosec}^2 l) \qquad (10)$$

When $l = 0$ this becomes

$$dm = dl\ (1 + \frac{n^2}{2}),\ \text{since}$$

$$\tan \theta = \theta + \frac{\theta^3}{3} + \frac{2\theta^5}{3.\,5} +,\ \text{and}\ \sec \theta = 1 + \frac{\theta^2}{2} + \frac{5\theta^4}{24} +$$

If $n = 3°,\ 1 + \frac{n^2}{2} = 1 + \cdot 00137 +$

If $l = 90°\ d\,m = d\,l$

or the elongation of the meridians on the chart increases as we recede from the poles.

The differential equation showing the elongation of the meridian on the projection of the spheroid may be obtained in a similar manner, but the ratio between the length of a portion of the meridian and the length of its projection may be obtained from the projection tables As we may consider the projected meridian to be straight for small distances, as 30′ of latitude.

Taking the latitude from 0° to 30′ and $n = 3°$, we shall have

$\delta y = \delta s + \delta\ \text{N} \cot l\ \text{versin}\ \theta = 55282$ metres + 76 metres.

The first number being the meridional arc, the second being the change in the tabulated value of y for a change of 30′ latitude.

$$\therefore\ \delta m = \delta y \sec \theta = \delta s\ (1 + \frac{76}{55282}) \sec \theta$$

Since $\theta = n \sin l$ is small, its secant can be taken equal to unity then,

$$\delta m = \delta s \ (1 + .00137 +)$$

The preceding formula (10) for the sphere gives for lat. 1°

$$d\,m = d\,l\,(1 + .00137 +)$$

So within the limits we have assumed there is no appreciable difference between the projections of sphere and spheroid so far as the elongation of the meridian is concerned.

CHANGES OF ANGLES CAUSED BY PROJECTION.

122. Let A be the azimuth, made by any elementary line on the surface of the sphere with the meridian ; let dp be the element of the parallel ; dl the element of the meridian on the sphere, and dp' and dm these elements when projected, and let A′ be the projected azimuth. Then $d\,p = dp'$ since the parallels are projected with their true lengths.

$$\tan \text{A} = \frac{dp}{dl}, \quad \tan \text{A}' = \frac{dp'}{dm}$$

At the equator where there would be the greatest difference between the angle and its projection this becomes

$$\tan \text{A}' = \frac{2}{2 + n^2} \tan \text{A}$$

Let $\dfrac{2}{2 + n^2} = a$, then the azimuth that will be most affected by projection will be such an one that

$$\tan \text{A} = \frac{1}{\sqrt{a}} \tan 45° = \frac{1}{\sqrt{a}}$$

If $n = 3°$, $a = .99863 +$, and A $= 45°\ 01'\ 10''$ and

tan A′ $=$ tan $(45°\ 01'\ 10'') \times a$, or A′ $= 44°\ 58'\ 49''$

Hence it follows, that the maximum change caused by projecting azimuths will not exceed two and one third minutes, if n does not exceed 3°

ANGLE BETWEEN MERIDIANS.

123. The angle between any two meridians on the globe in lat l is found by the equation,

$$\sin \tfrac{1}{2} A = \sin l \sin \tfrac{1}{2} n.$$

On the projection when n is not greater than 3°

$\tfrac{1}{2}\theta = \tfrac{1}{2} n \sin l = \tfrac{1}{2}$ angle between projected meridians nearly.

Since $\sin \tfrac{1}{2} n = \dfrac{n}{2} - \dfrac{n^3}{48} + \&c.$ and $\sin \tfrac{1}{2}\theta = \dfrac{\theta}{2} - \dfrac{\theta^3}{48}$

We have, neglecting higher powers,

$$\sin \tfrac{1}{2} A = \frac{\theta}{2} - \frac{n^3}{48} \sin l$$

$$\therefore \sin \tfrac{1}{2} A = \sin \tfrac{1}{2} \theta + \frac{\theta^3}{48} - \frac{n^3}{48} \sin l$$

$$\therefore \sin \tfrac{1}{2} A = \sin \tfrac{1}{2} \theta + \frac{n^3 \sin^3 l}{48} - \frac{n^2}{48} \sin l$$

$$\therefore \sin \tfrac{1}{2} A = \sin \tfrac{1}{2} \theta - \frac{n^3}{48} \sin l \cos^2 l$$

The second term 2d member will have nearly its maximum value when $l = 35°$

$$\text{Sin} \tfrac{1}{2} A = \sin (57' - 22'') - .000001151$$

or A will differ from θ less than $\dfrac{5''}{10}$ when $n = 3°$

PROJECTION OF SHORTEST DISTANCE BETWEEN TWO POINTS.

124. The co-ordinates of the projection of any point being given by

$$x = N \cot l \sin \theta$$

$$y = N \cot l \operatorname{versin} \theta + S$$

$$\theta = n \sin l$$

by obtaining the latitudes and longitudes of different points in the arc of the great circle, (Geodesic line) between two points, we can compare the co-ordinates of the projection of these points with the co-ordinates of a straight line, which either passes through the projections of two extreme points of the great circle, or cuts the middle meridian at the same angle as the great circle, which angle will be the same on the projection as on the spheroid, as since the elements about points in the middle meridian are projected with their true lengths, the equality of angles must be preserved.

A Geodesic line on the surface of the spheroid is the shortest distance between two points. It corresponds to the Great circle arc on the sphere.

Suppose through a point on the spheroid whose latitude is l, the surface of a sphere be described with radius N, the surface of this sphere will cut the surface of the spheroid in the parallel through the point, since the centre of the sphere lies in the minor or polar axis, see fig 4, as do also the centres of parallels, and since the radii of the parallels of spheroid and sphere, passing through the point are both N cos l, the two parallels must coincide. Planes passed through the axis will cut sphere and spheroid in the meridional arcs; and longitude from the middle meridian will be the same for both figures. since longitude is the angle between these planes. Any plane passed through the normal, will cut the surface of the sphere in the arc of a great circle and the spheroid in a line which for three degrees will be sensibly the geodesic line. Within that limit these lines may be considered equal. (See Puissant Traité de Geodésie.)

By assuming the angle this plane makes with the meridian, we can easily determine the latitudes of the intersections of the great circle arc with different assumed meridians, by the formula of spherical trigonometry. We should thus get the departure in latitude of the great circle arc at the various meridians from the latitude of the origin (or from the parallel of the point in which it cuts the middle meridian.)

This departure, if expressed in linear units, would give in those units the departure of the geodesic line from the same parallel on the same meridians of the spheroid.

If the change in latitude on the spheroid is desired, we must remember that latitude on the elliptical meridian at any point is measured on a circle, whose radius is R_m; and though this radius is constantly changing, it may be considered constant within small limits, especially if we take its value always for the middle latitude; therefore, the change in latitude on the sphere will be to the change of latitude on the spheroid as $R_m : N$.

This principle may be thus expressed :—

The same distance measured on arcs of different curvature will subtend angles, inversely, proportional to the radii of the arcs; therefore, after obtaining the differences of latitude on the sphere, we may find the differences on the spheroid by multiplying by $\frac{N}{R_m}$, which ratio can be obtained from the projection tables. We shall then have the latitudes and longitudes of the points in which the great circle cuts the meridians on the spheroid. These points can then be projected by the projection tables, and we can then see how the line drawn through the projections compares with a straight line. For instance, suppose the plane through the normal is perpendicular to the meridian.

Let the latitude of origin be 45°N, the assumed meridians 30′, 1°.00, 1°.30′, 2°.00′, 3°.00′.

The departures from the parallel to the Sd. will be found for the sphere 3″.92, 15″.72, 35″.33, 1′. 2″.83, 2′. 21″.45

For spheroid 3″.933, 15″.773, 35″.449, 63″.041, 141″.93

$-\ S = R_m \sin 1'' \times 3''.9$ &c.

$-\ S = -\ 121.41,\ -\ 486.86,\ -\ 1094.2,\ -\ 1945.9,\ -\ 4380.8$

in metres.

The values of $y = N \cot l \text{ versin } \theta$ can be taken from the projection tables.

y = 122, 486, 1095, 1946, 4378, in metres.

In this case we shall have for the ordinates of the projections of the points $y = N \cot l \text{ versin } \theta - S$ when we refer these points to a perpendicular axis through 45°, whence the ordinates are

+ .59 — .86 + .8 + .1 — 2.8 metres.

As the values of y, in the projection tables, are only given to the nearest metre, these results only determine the fact that the projection of the geodesic line perpendicular to the meridian in latitude 45° does not depart from the line perpendicular to the middle meridian, an amount equal to a metre, until it has reached beyond the meridian of 2°; and its departure on the meridian of 3° is less than three metres. As this amount would not be appreciable on any chart which would have an amplitude of six degrees longitude, we may say that the projection of the great circle is a straight line within the limits we have taken. Since the same would be true of the line and great circle on the other side of the middle meridian, it follows that the projection o the great circle passing through two points on the same parallel whose longitudes are 3° E and 3° W, will be a straight line.

In a similar manner it may be shown that when the great circle makes an angle of 45° with the meridian, the projection of the geodesic line of the same *length* as before considered sensibly coincides with a straight line. The length of the lines considered is about one hundred and forty miles.

In latitude 80° the projection of the point in which the geodesic line, perpendicular to the middle meridian, cuts the meridian 4° from it, falls only 6.6 metres from the line drawn on the projection perpendicular to the meridian at latitude 80°. The length of the geodesic line from middle meridian to point of intersection in this case is about 48 miles.

In addition to the greater uniformity of scale the above examples show the advantages the polyconic possesses over Mercator's projection for surveying purposes.

Instead of the great circle arc perpendicular to the meridian in latitude 45° being projected sensibly as a straight line, on the Mercator projection, it will be projected as a curve which in longitude 3° will deviate from the straight line perpendicular to projection of meridian nearly 4381 metres × sec latitude.

In latitude 5° the projection of the perpendicular great circle arc sensibly coincides with the perpendicular straight line for 3° of longitude.

GENERAL CONCLUSIONS.

125. We see from what has been shown that the ordinary polyconic projection, when its amplitude in longitude does not exceed 3° from the middle meridian, has the following properties. It is well adapted to all parts of the earth, but best to the polar regions.

The meridians make practically the same angle with each other, and with the parallels as on the sphere.

Angles will be projected with but little change.

The great circle (geodesic line) will be projected as a straight line practically equal to itself.

RECTANGULAR POLYCONIC.

126. The rectangular polyconic of the British War Office is a projection where the parallels are described as before, but as the meridians

must cut them at right angles, they are no longer projected in their exact length, and θ is determined from the new condition imposed, that of rectangularity.

For particulars as to this projections see *Traite des Projections des Cartes Geographiques. Par A. Germain. Arthur Bertrand, Editeur, Paris.*

CHAPTER XI.

RUNNING SURVEY.

GENERAL METHOD OF RUNNING SURVEY—PRECAUTIONS AS TO OBSERVATIONS—PLOTTING AND BALANCING THE WORK.

GENERAL OBSERVATIONS.

127. Could the reckoning of a ship be *accurately* kept as she ran along a coast we could evidently make a very good chart of it, simply by taking accurate bearings of various points on the shore line and noting the time. The track of the ship would be a base line, the intersections of the bearings would fix the positions of the shore line. Having determined the latitude and longitude accurately by observation at intervals of forty or sixty miles, the intervening points could be plotted by plane sailing. Now we can determine the bearing of any terrestrial object from the ship with great accuracy by the astronomical method, but owing to currents, lee way, and difficulties in steering, can not depend on the accuracy of the track base. Neither can we multiply our astronomical stations for the determination of latitude and longitude, for each such determination has errors peculiar to the method used, and by reducing the interval between astronomical stations we should soon reach a condition of affairs when the distance between stations could be more accurately determined by the patent log than by observation. In other words the absolute error in the determination of the length of the astronomical base (which is in reality the distance between the astronomical stations) is no greater, necessarily, when such base is a thousand miles long than when it is five miles long; but evidently the proportional error is very much

less. We should therefore seek to reduce the errors of positions determined by astronomical observation to a minimum, therefore the observations should be made ashore, with the transit and zenith telescope if possible, otherwise reflecting instruments and the artificial horizon may be used, in which case the Pistor and Martin prismatic circle may be used, as its power of measuring angles of 180° with accuracy, would enable us to observe bodies near the zenith, thereby reducing the probable error of refraction.

Landing will often be difficult, sometimes impossible, in which case the position must be determined by altitudes from the sea horizon.

CARE AS TO OBSERVATIONS.

128. We can not hope for very accurate results from such observations and should be careful to take the following precautions. The vessel should anchor so her postion may be fixed by angling on shore marks, and so as to have as great an extent of clear sea horizon as possible. If possible, time positions should always depend on observations at equal altitudes on opposite sides of the meridian, latitude positions on equal altitudes on the meridian north and south of the zenith. The effects due to errors in the dip-tables should be checked, either by simultaneous fore and back observations, or by measuring the angle between opposite points of the the horizon with a prismatic circle. This angle will evidently be 180° + 2 dip.

Owing to obscurity of the sea horizon at night we shall seldom be able to obtain equal and opposite meridian altitudes for latitude.

The above rules for observations, save as to the dip, will apply to observations taken ashore.

ACCURACY OF WORK.

129. As the navigator will not, in general, be able to fix his position with as much accuracy as we have fixed the positions of our stations, even when we have been prevented from landing, it follows, if we can fill in the detail of the coast with accuracy (that is so that the absolute error in the distance between two astronomical stations may be distributed proportionally along the coast between them), we shall have a chart whose accuracy of shore line and hydrography will be equal to any test that he can give it. In other words, we may make an accurate chart of a coast without landing upon it if we can elimi-

nate the ordinary errors of the run of the ship between stations. We can almost entirely do this when the coast presents a succession of views of prominent objects by the following method:—

EXPLANATION OF METHOD.

130. In the sketch, Plate VII., O represents the station occupied, and whose latitude and longitude are accurately determined by observation.

The ship leaves her anchorage at O, leaving a party ashore; she proceeds to some point A, in as direct a course as possible. The distance OA should be such as to form well conditioned triangles with the directions of the prominent objects ashore, as *a*, *b*, *c*, *d*, *r*, &c.

On reaching A she may anchor or not; in either case, at a given signal, simultaneous angles are observed at A, between O, and the above points; at O the observers measure the angles between the points and A; an astronomical bearing of one of the points is taken at the same time either at O or A.

The ship then goes to B, steering on the range *c d*. The shore party proceed along the coast by boat, angling on the points *a*, *b*, *c*, *d*, *r*, &c., and sounding; the ship sounds constantly, noting the times. On arriving at B, the ship takes the astronomical bearing of any two points, as *a* and *d*, or measures the angles between three points as O, *a*, and *d*; at the same time angles are measured between any one of these points and some new objects which have just come in sight, these are *e*, *r'*, and *f*.

The vessel then goes to C, steering on a range if possible; at C she takes astronomical bearing of *d*, and measures the angle between *c* and *d*, this will fix her position in reference to those points. At the same instant the angles between *e*, *r'* and *f*, and *d*, or *c* are measured, these will fix *e*, *r'*, and *f*. At the same instant the angles made by the points *r''*, *g*, *h* which have now become visible are measured.

The ship then goes to D, fixes her position by astronomical bearing of a known point and the angle between it and another known point, and by simultaneous angles on the points *r''*, *g*, *h*, fixes their positions, which in turn will serve to fix new positions of the ship, and the places of other points which come in sight as the ship goes towards O'. On arriving off O', the ship anchors, sends a party ashore, and O' is fixed by observations on the ship and fixed points. The latitude and longitude of O' are accurately determined by observation.

It would have been well if a reconnoisance of the coast, between O and O′, had been made the day before the survey commenced. Compass bearings of different landmarks, as well as sextant angles and the patent log would have enabled us to make a sketch of the coast which would have been of great assistance in the survey. The astronomical positions of O, and O′, would have been determined by sea observations, and we could then have selected a middle meridian of a polyconic chart so that O and O′ could be projected having equal longitudes in reference to it. O and O′ should never have a greater difference of departure than sixty miles on account of the projection, but the difference of latitude between the points may be no more than ten miles, if the positions are determined by good observers with the zenith telescope. But we must remember that the great trouble in the survey will be in landing. It will most always be inconvenient and often dangerous, therefore, we must take advantage of favorable opportunities, such as good weather, sheltered coves, &c., to determine our shore positions, considering sixty miles as the extreme limit that should exist between shore stations, but preferring that they should be thirty miles apart.

The amount of coast surveyed each day will depend entirely on the detail; the more lines run and soundings made the less will be the progress down the coast.

The surveyor should remember that satisfactory results will be much more apt to follow careful work, as the additional data will give more checks. Therefore, often the ship may be several days in going from station to station. She should anchor every night and check her position by observation whenever it can be done under favorable circumstances.

PLOTTING THE WORK.

131. Let us again refer to Plate VII.

Before leaving O, we should have determined, either from the reconnoissance or other sources of information, what would be the probable extent and direction of coast to the next shore station. A polyconic chart to include the coast, scale depending on the extent of coast to be projected, the greater the extent of coast the smaller its scale, should have been constructed, and O projected with its proper latitude and longitude. The middle meridian should be assumed half way between O and O′.

The distance O A in units of the chart scale should have been determined either by angles from a measured base at O, mast head angle, or by the patent log. The vessel might have run out on a range, to a point where it was cut by another range, hauling in and reading the patent log at the instant of crossing the 2d range, when angles are taken, then turning and going back on first range, putting over the log at time of crossing 2d range, she returns to point of departure determined by a buoy; the mean of patent log readings will give in connection with distance from buoy to O the distance O A.

The distance O A, can then be plotted with its proper azimuth, and the points *a*, *b*, *c*, *d*, *r*, determined by the angles from O and A, being plotted by protraction.

132. We supposed the ship went directly from A to B. Evidently as she can now fix her position at any moment by angles, it is not necessary that she should proceed to any *particular* point. Her movements may now be governed by two considerations only.

1st. The triangles formed by the points ashore and her positions should be well conditioned, and angles to all new points should be well conditioned. This rule should always apply, whether the survey is to be minute or not.

2d. The number of the lines beyond those necessary to fulfil the first condition, will be proportional to the time to be devoted to the work.

In carrying out the 1st condition it is evident a preliminary sketch of the coast would be very useful.

B, C, D, E, F, together with the various shore points, are plotted either by astronomical bearing of a known point and the angle between it and another known point, or by the three-point problem.

ASTRONOMICAL BEARING.

133. The astronomical bearing is preferred, as by plotting from the particular meridian of the place where it was observed; as we change our longitude, we allow for the convergence of the meridians. Therefore, if the sun is over the land, the eye must be placed as low as possible and the observation corrected for the dip of shore line, or time azimuths should be used. As the time can be very correctly determined twice a day without landing, and as the change of longitude is known, and as the chronometers will be of the best, the time

azimuth will probably give the best results. It should always be used in preference to the other as the sun approaches the meridian.

If possible it would be best in practice to determine the position of prominent land marks by astronomical bearing, in the morning and evening when the sun is low. These objects can give good positions for the ship and consequently to new objects as they come in sight, and we can thereby avoid the use of the sun for determination of bearings when its altitude becomes so great as to give unsatisfactory results.

From art. 279, Coffin, we see that the problem becomes indeterminate when the great circle, passing through sun and object, is perpendicular to horizon.

As this problem is discussed in the work referred to, it will not be here; but the attention of the surveyor is particularly called to these facts.

As his results depend entirely on the accuracy with which the astronomical bearing is determined, he should understand the problem thoroughly; he will then select the method that will give the best result, and when the probable error exceeds what he has determined to be the limit, should interrupt his operations till a more favorable time.

In all that has preceded, we have only mentioned the sun, in connection with the astronomical bearing, but simply because generally it will be used. Often the moon and planets will give better results and should be used in preference to the sun.

BALANCING THE WORK.

134. When we plot O′ as determined by observation, we shall most likely find that its position differs materially from the one given by the triangulation from O. We must therefore balance the work to make the two positions co-incide.

From chap. X we have as follows :—

The straight line OO′, on the Polyconic chart, within the limits assumed, is the projection of the great circle through those points. The lines OA, AB, BC, &c., are likewise projections of great circles through A, B, C, &c.

The angles between these arcs are not sensibly altered by projection; neither do the projections sensibly differ in length from the arcs themselves. We may therefore balance the work in a similar manner as we would in plane surveying.

Draw a straight line through O O′, and measure the inclinations of all the lines OA, AB, BC, &c., to the line O O′. Let these inclinations be I, I′, I″, &c., then the diff. lat.* for AO for instance = A O × cos I. The departure = AO × sin I, and so on for the other sides.

The sum of the latitudes should equal O O′ and the sum of the departures should be zero if the determination of O′ by observation and triangulation agreed.

Each partial diff. lat. and departure will be corrected by the quantities obtained as follows :—

[AO + AB + BC + &c.] : AO = Total error in lat. : correction for lat. of AO

[AO + AB + BC + &c.] : AO = Total error in dep. : correction for dep. of AO

From the new values of the diff. of lat. and departure for A O, A B, B C, &c., new values for them and I, I′, I″, &c., are to be obtained from the traverse table. The work is then to be replotted, the new lines A O, A B, B C, &c., being considered the bases, and the angles to the shore points being taken the *same* as before. These lines A O, A B, B C, &c., are to be plotted by their inclinations to line through OO′, and their correct lengths taken from the scale. The points O, O′, are to be taken (in their proper latitudes) so they have equal longitudes from the middle meridian. The shore line would be sketched in, and soundings not fixed by angles would be put in, in the usual manner, spaced between principal points by proportion.

135. In balancing the work, we have assumed that the errors are proportional to the lines O A, A B, &c. Of course if O A was not correctly measured, the work could not close.

Often, however, certain points as D, for instance, are determined by small angles, and may be assumed to be erroneously determined for that cause, and it might be well to assign an arbitrary compensation to D, or to say that compensation was directly proportional to the length of a line and inversely proportional to the angle that determined its extremity considering the right angle the greatest. The patent log should be used and its readings would act as checks. For instance, if the patent log agreed well with the distances B C, D E, but disagreed with C D, as determined from the chart, we should be justified in giving greater compensation to C D than to the other lines.

* The term diff. lat. and departure are only used as they concisely express the idea. We really get the projections of the lines on O O′ and on a perpendicular to O O′.

CHECKING POSITIONS.

136. Another way of checking positions will be by determining the vertical heights of prominent points, and using these heights afterwards to check distances. For instance, if the height of *e* had been determined at B, by taking its vertical angle, the distance deduced from the vertical angle of *e*, as seen at D, would serve as a check on the position of D.

137. It would always be well to avoid small angles by the following precaution. On approaching a point, such as D, where it is evident the angles measured between known points will be small, a boat should be sent to some point x; the boat fixes her position with good angles, at the same time the ship angles on the fixed points and the boat.

BOAT WORK.

138. But little has heretofore been said about the work of the boat or boats during the survey. Evidently the ship can do all that has been described with but little assistance from the boats, but as we make the survey more minute, so will the boats become more useful. If all the lines run by the ship were similar to O A, and A B, off and on the coast, making a series of triangles as nearly equilateral as possible, the boat would do good service by sounding the interior lines as O B, or the bases of the triangles. The boat and ship would communicate at B, when the points to be angled on for the next triangle could be named. In other words, the special work of the ship would be to run the outside lines of soundings, fix and name the prominent shore stations, or objects, which she can better do, as she changes her position more, and will more often have a clear horizon. The boat, by communicating with the ship, would obtain a rough sketch of the coast and the positions and names of prominent marks, and would run her lines of soundings by angling on these marks.

After the final adjustment on the projection of the principal shore points, the various lines run by the boat may be plotted by means of the protractor.

GENERAL RULES.

139. The following general rules may be given.:—

Avoid as much as possible angles less than 30°.

Note transits and tangents.

When the position of any point is determined, determine its height, always correcting for the terrestrial refraction and dip of shore line and curvature of earth.

Steer as much as possible on ranges, *noting* the compass course, the track of the ship will then be a straight line and the soundings will be more accurately plotted. If you cannot get a range, keep a point on the same compass bearing, to effect the same object. The length of the lines of triangulation made by the ship will depend on the distinctness of the objects viewed, they would generally be less than five miles long.

Keep three known objects, subtending angles over 30°, constantly in view; whenever a new point appears, whose direction makes an angle of 30° with direction of ship's track, determine its direction and position of ship. Again determine its direction in reference to known points, and change course before the angles between known points become 30°.

Angles between two known points and an astronomical bearing of one of them will fix the position of the ship. As there is no ambiguity attached to this method as is frequent when three points are angled on, it must be constantly used. As good results cannot be expected to be obtained by inexperienced observers, however able they may be, as the success does not depend so much on individual ability as on concerted and well directed action, it follows that the observers should be thoroughly drilled before the actual work of the survey commences. This can best be done by going over the lines several times in the immediate vicinity of the extremity; comparison of results will expose errors and make the method familiar, and demonstrate the degree of confidence with which to regard it. A mountain, far inland, can be fixed in position by astronomical bearings from the shore stations; its bearing at intermediate points will check those points. The rule in regard to angles applies here.

No point should be fixed by lines making a less angle than 30°, if the survey furnished larger angles. In other words, rather than fix the position of a point from a short base line, expand your base line by the triangulation, and consider the point fixed by the best angles. If the coast presented successive views of prominent objects, such as mountains, rocks, cliffs, &c., visible at a great distance, as fifteen or twenty miles, three of which would always be in sight, the track triangles must be larger and would enclose a smaller triangulation to give the details of the coast. The rule in reference to angles will determine the length of sides.

Whenever the ship anchors, currents should be observed, and the rise and fall of the tide by pressure gauge.

The principal points determined each day should be plotted immediately, and a sketch made for the boats and general reference.

RESULT NOT AFFECTED BY ERROR IN ORIGINAL BASE LINE.

140. It will be seen from the preceding pages that the positions of points are made to depend on the distance between the shore stations as determined astronomically. These distances from point to point are independent of the original base line OA. Distances by patent log and compass courses are only used as checks.

SPECIAL CASES.

141. If at either astronomical station a shore triangulation be established, as a harbor survey, whence, by an expanded base, the positions of peaks are determined by well conditioned triangles, then the various positions of the ship, fixed from these peaks, *may* be considered correct, and the triangles about the station plotted without change.

For instance, Pl. VII. If ,by an accurately measured base at O, the positions of *a*, *b*, *c*, *d*, be determined, the positions of A, B, C, determined from them as well as *r*, *r′*, *e*, *f*, may be considered correct and plotted without change. Compensation must then be made between C and O′ in the manner before described.

If at O′ there also be a measured base whence the positions of peaks are determined, that portion of the triangulation from either point towards the other that depends on well determined points may be plotted as correct. As the two triangulations, when brought to the same intermediate point, will not agree, a certain portion of the triangulation about such point should be balanced, as before described. Should any portion of the coast between astronomical stations be without natural signals, the triangulation would be carried from the origin as far as may be considered accurate (till angles between known objects measure 15°), and further positions fixed by dead reckoning.

If, before reaching the other astronomical station, other signals come in sight the triangulation should re-commence.

We can then plot the triangulation from each station and connect by dead reckoning, balancing as before or giving greater compensation to distances determined by dead reckoning.

Cases may arise where the triangulation must be connected with the origins by dead reckoning. Even in such cases the chart will be of the greatest use to the navigator ; for should he plot his position on it by cross bearings of any of the natural signals used in the survey, such position will give the correct *bearing* of dangers, though the distances as given by the chart, will most likely be incorrect.

Correct bearings are of the utmost importance in a survey. Distances, are of secondary importance, since, owing to currents, the speed of the ship is approximate.

PROJECTION PLACED ON PLOT OF TRIANGULATION.

142. If working in low latitudes or in the direction of the meridian in high latitudes the convergence of the meridians can be neglected.

We then plot the triangulation on plain paper, and afterwards put the polyconic projection on it so that the extreme points, which we shall call O and O′ may be projected in their proper latitudes with equal longitudes from the middle meridian. By this means compensation is made to all the lines of the triangulation.

To do this, plot on a polyconic chart the positions of O, O′ as determined by observation, their longitudes from middle meridian being equal.

Draw a straight line through these points, determine, by measurement, the length in metres of its representative on spheroid, which call a. Measure the angles this straight line makes with the co-ordinates of the points.

On the plain paper, having the triangulation on it, draw a straight line through O and O′, measure its length in metres. Call this b, then $\frac{b}{a}$ represents the co-efficient of the scale of the desired projection. By laying down the co-ordinates of O, O′, as taken from tables and reduced by co-efficient of scale, making the same angles with line through O, O′ as were measured on the chart, we fix the middle meridian and points in which parallels through O and O′ cut it. Other parallels and meridians can then be drawn if desired.

For further information in regard to running surveys, the student is referred to Hydrographic Office publications, Nos. 4 and 36, and "Directions as to the Running Survey of the Mexican Gulf Coast." Also to the directions given in the Deck Board and Angle Book issued by same office.

CHAPTER XII.

THE PORTABLE TRANSIT—GENERAL DESCRIPTION—ADJUSTMENTS—POINTING—DETERMINATION OF VALUE OF TURN OF AZIMUTH SCREW—EQUATORIAL INTERVAL OF THREADS—REDUCTION TO MERIDIAN—DETERMINATION OF CONSTANTS a, b, c.

GENERAL DESCRIPTION. (See PLATE.)

143. The smaller of the portable transit instruments used by the U. S. Government, built by Wurdemann, of Washington, has a telescope of 26 inches focal length with 2 inch aperture and magnifying power of about 40. The frame made light and very stable consists of a horizontal portion fitted with two levels at right angles to each other and three foot screws by means of which it may be levelled. At the extremities of this horizontal portion are hinged the two standards fitted with braces, which set up firmly by means of screws, against two studs in the lower part of the frame. On removing these screws the braces fold in upon the standards and these latter down, so that the entire frame may be packed in a box together with the telescope convenient for transportation.

To the upper end of each standard is attached a Y, or the carefully manufactured bearing for the extremity of the horizontal axis. One of these Ys is fitted so that a fine motion screw with micrometer head gives an accurately measured motion in azimuth and the other so that a similar screw gives a similar vertical motion. Supports are arranged by means of which lamps may be used to illuminate the field of view and the ordinary stops for the clamps are fitted to each standard.

To the telescope tube are attached the two conical supports by their larger ends, the outer extremities being ground down carefully to form the cylindrical pivots which rest in the Ys. The axes of these supports, in the same straight line perpendicular to the line of direction of the tube, form the rotation axis of the instrument. The supports are made conical to prevent flexure, and hollow in order that light from the lamps may be admitted through them to the reflector in the telescope tube, which serves to render the transit threads visible when observing faintly illumined bodies at night. Near the extremity of one of the axial supports is the clamp with thumb screw, fitted with lever and spring, which latter, in connection with thumb screw of stops on the standards, enables us to give slight motion of the line of sight in a vertical plane for the more accurate pointing of the instrument.

On either side of the telescope tube, between the eye piece and axis, are the two finders, each of which consists of a graduated circular scale about 4 inches in diameter permanently attached to the tube, and a spirit level secured to an arm, which revolves about the centre of the circle, fitted with vernier, stop and tangent screw, by means of which single minutes may be read on the scale. One of these finders may be set to read altitudes, and the other zenith distances.

The telescope consists of the tube proper, in which is placed the object glass, and the eye tube, which contains the diaphragm or wire plate and the eye piece; the eye tube is moved out and in by a rack and pinion so that having adjusted the wires to the focus of the eye piece they may easily and quickly be brought to the focus of the object glass. Another rack and pinion carry the eye piece to the right and left so as to bring its optical centre opposite each thread in succession as a star crosses the field. The diaphragm or wire box contains the reticule which consists of nine parallel transit threads with three others at right angles; there is also a movable thread and micrometer-headed screw carrying it—all so arranged with strap and clamp screw that they may be revolved in any direction to make the threads vertical. The wire plate has also a small bolt attached to it carrying the wires, when vertical, to the right or left for the purpose of adjusting the line of sight over the middle vertical wire perpendicular to the axis of rotation. A diagonal eye piece is furnished for observing stars at considerable altitudes.

The striding level, with its accurately divided scale, is filled with ether, and furnished with a chamber to regulate the length of the air bubble and also with adjusting screws for regulating the reading of the bubble with reference to the level supports.

ADJUSTMENTS.

144. Having selected the position for mounting the instrument, the stone pier, block of wood, or whatever is to be used as a support, should be placed as firmly as possible on its foundation and made level. (For field work it is recommended to use the trunk of a tree cut off at a convenient height.) Place the frame of the transit with the two standards in the east and west line as nearly as can be determined by means of the compass and its corrections for deviation and variation. By means of the foot screws level the horizontal portion of the frame, and set up the standards by means of the braces, taking care to set the screws connecting them with the frame-studs well taut. See that the Y adjusting screws are in the middle of their position and then mount the telescope with the pivots in the Ys ; adjusting the lever of the clamp in the stops.

145. LEVELLING.—To insure the motion of the axis of collimation * in the plane of a vertical circle it is necessary that the axis of rotation be made horizontal ; to set it in this position apply the stride level with its supports on the pivots. If the level is adjusted so that the axis of the tube is exactly parallel to its supports, it will only be necessary to bring the bubble exactly in the centre by means of the Y adjusting screw giving vertical motion. It is not well however to trust the level without reversal, consequently shift it end for end, and if the bubble in the new position is not exactly in the centre of its reading, make it so by turning the Y adjusting screw to correct one half the error and the small screw for adjusting the level tube with regard to its supports, to correct the other half. Again reverse the level, and if the bubble remain central, the axis is horizontal. If not, repeat the above operation until the bubble remains central in both positions of the level, this being the test for the correct position of the *horizontal* axis.

VERTICALITY OF WIRES. COLLIMATION.

146. Direct the telescope to some small, distant, well defined object † and bisect it with the middle wire. Elevate and depress the telescope on its axis and observe whether the object still remains bisected by every part of the wire ; if not, loosen the strap which holds the mi-

* The axis of collimation is that imaginary line passing through the optical centre of the object glass perpendicular to the axis of rotation and in the same plane.

† The cross wires of a collimating telescope or theodolite may be used with the best results for the distant object.

crometer box and diaphragm, and turn until the central wire is vertical, evidenced by the above test. Next make the line of sight over the middle vertical wire correspond with the axis of collimation—called the adjustment in collimation. Observe that the distant object is still bisected by the middle vertical wire and then reverse the telescope by lifting it carefully out of the Ys, and, replacing it with the axis shifted end for end, point again to the same object, moving the telescope only in the vertical plane. If the object be bisected in this new position, the adjustment is correct; if not, by means of the screw bolt holding the diaphragm, move the middle wire half the distance from its new position toward the middle of the distant object, then, by means of the Y azimuth screw, bring the wire into position bisecting the object, and reverse again. If the above condition be fulfilled in this new position, the line of sight corresponds with the axis of collimation; at all events, the operation of correcting half the apparent error by means of moving the diaphragm, must be performed until the middle wire bisects the distant object in both positions of the axis.

It is evident that the collimation adjustment in no way depends upon the direction of the telescope, but, as it is nearly always possible to find some small distant body in any direction, it is not generally made until after the axis has been levelled approximately in the east and west line and the wires made vertical.

147. The final adjustment is that in azimuth to insure that the plane of the line of sight is in the plane of the meridian; this can be done only by observation.

TO POINT THE TELESCOPE.

148. This is done by means of the finder. That finder which reads altitudes should have the zero of its scale in such position that when the vernier reads zero the axis of the level should be parallel to the line of sight of the telescope—if the vernier be so set and the telescope then revolved until the bubble is in the middle of the tube, the line of sight will be horizontal. To point to any star whose altitude is H:—set the vernier to that reading and revolve the telescope until the bubble be central. It is evident that the angle made by the line of sight with the horizontal line is equal to the angle read from the scale, in this case equal to H, generally the computed meridian altitude of a star.

In order to find any index error or to set the scales of the finders, the following method is used. By means of the eye direct the telescope to any known star at time of culmination and clamp it; bring the vernier arm level by means of the bubble and, to find index error, compare the reading of the scale with the computed reading; the difference will be the error required. To set the scale proceed as before, loosen the screw holding the scale in position, turn it until the indicated reading be the same as the computed, and secure it. Compute the zenith distances of stars at time of culmination by means of the general formula for the Latitude, $L = z + d$, where z is the meridian zenith distance and d the declination. Then $z = L - d - r$ for star south of zenith, and $z = d - L - r$ for star north—taking care to use the proper sign for d and in case of a sub-polar star to use $180° - d$ instead of d. r is the refraction which may generally be neglected. The altitude $H = 90° - z$.

TO MAKE THE ADJUSTMENT IN AZIMUTH.

149. Since the rotation axis has been made level and the line of sight perpendicular to that axis, it is evident that, when vertical, the line of sight will pass through the zenith and will therefore cut the meridian at that point, however far removed it may be when turned down to the horizon. If now we can make the line of sight cut the meridian when pointed in any other direction, the plane in which it moves will coincide with the plane of the meridian. A star *in* the zenith will appear to cross the middle vertical wire at the time it actually crosses the meridian, but such will not be the case with a star near the horizon unless the azimuth adjustment is exact. If the low star, crossing from east to west, passes the middle wire earlier than the computed time of transit, the plane of the instrument deviates to the east, if later, the plane deviates to the west; in either case the error must be corrected by the azimuth screw, until stars at all altitudes indicate the same amount of clock error. Practically take two stars, one in or very near the zenith, the other as near the horizon on the same side of the zenith as possible, the stars to differ but little in right ascension. Note the clock time of transit of the zenith star and compare this time with its right ascension: the difference will be the approximate clock error. This error, applied with its proper sign to the right ascension of the second star, will give the clock time of its transit; point the telescope to the second star and by means of the azimuth screw, if necessary, bisect it with the middle wire at the com-

puted time of its transit. The instrument will then be very nearly in adjustment, but by repeating the above operation it may be brought more nearly, and, if there be time, the adjustment may be made exact by repeated observations and azimuth screw turnings. The error may be found, however, and the adjustment made much more quickly, by the following method, the mathematical part of which will be explained later. Take two stars as before differing but little in right ascension in order that the interval between the times of observations may be correctly measured, *i. e.*, unaffected by whatever rate the clock may have, and differing greatly in declination. Note the time of upper culmination of each star and mark the interval between, positive when the lower star precedes the higher, negative when it follows; mark the difference between the computed right ascensions of the two stars in the same way. If these two quantities are exactly the same, the line of sight is exactly in the meridian; if not, the difference between them points out a deviation, in northern latitudes, to the east of south when positive, and to the west when negative, the amount of deviation in time expressed by the formula

$$D = \delta \cos d \cos d' \operatorname{cosec} (d - d') \sec L$$

and 15 D equal to the deviation in arc,

δ being the difference of right ascensions less the interval between the times of transit; d' and d, the declinations of the higher and lower stars respectively, and L, the latitude.

150. Knowing the amount of deviation from the meridian, if we determine the value of a revolution of the azimuth screw, it is evident that we can at once bring the instrument into adjustment in azimuth by giving the screw the requisite number of turns. To determine this value, note the time of passage across the middle wire of an equatorial star, turn the screw quickly through a complete revolution so as to bring the line of sight to the westward of the meridian, and then note the time of its second passage over the middle wire, the interval in time v between these two passages, reduced to the horizon by increasing it in the ratio of the cos. of the altitude to one, will be the value required. $V = v \sec H$ or since H, the altitude of an equatorial star, is equal to the complement of the latitude, we shall have V in time equal to v cosec L, and 15 V the value in arc. If any other than an equatorial star be used, v, the noted interval, must be reduced to the equator * and its value then substituted in the above.

151. The following modification of this method will give better re-

* ($v = $ v cos d.)

sults. Determine the clock error by three or four stars of nearly the same declination, and following each other in quick succession so that chances of error due to the rate of the clock may be eliminated, turn the azimuth screw through several revolutions and take as many of the same and other stars as needed; then the difference of the clock errors, shown by the two sets of observations, divided by the number of revolutions given the screw, will be the value of a revolution which is to be reduced to the horizon as before.

Divide the error of the instrument in azimuth D thus found, by V, the value of a single turn, both in time, and the result will be the number of turns to be given the screw in order to bring the plane o the instrument to the meridian.

REDUCTION TO THE MERIDIAN.

152. If the foregoing adjustments are *perfectly* exact, that is, if the horizontal axis is exactly level, and in the true east and west line, and if the line of sight corresponds exactly with the axis of collimation, the great circle described by the line of sight will be the meridian. In practice, however, it is almost impossible to fulfill all these conditions exactly, and it becomes necessary to correct the times of observed transits for slight variations in these conditions; and these corrections are called the reduction to the meridian. The method for determining this reduction will be understood from the figure, in which the deviations are very much exaggerated in order that the method may be clear.

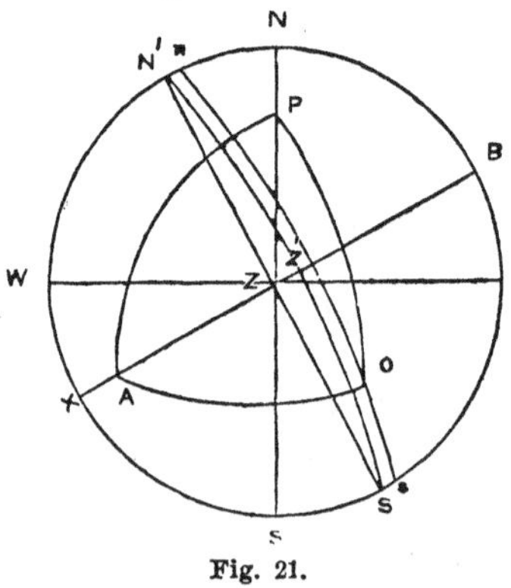

Fig. 21.

Let N E S W be a projection (stereographic) of the sphere upon the plane of the horizon; N S, the meridian; E W, the prime vertical; Z, the zenith. Suppose the instrument mounted so that the prolongation of its horizontal axis is represented by the line A B, meeting the celestial sphere at the point A; the deviation from the level will evidently be represented by the projected arc x A; this deviation is named b. The angle A Z W will evidently be the deviation of the axis from the E and W line, equal to P Z N′, the angle between the axis of collimation and the meridian, or deviation in azimuth, named a. Since the axis of collimation is that line perpendicular to the axis of rotation, it will describe the plane of a great circle of which A is the pole: let N′ Z′ S′ be that great circle, and let n O s be the circle described by the line of sight passing through the centre of the middle vertical wire, making the constant angle c with the axis of collimation, c being the collimation constant.

If the line of sight moves in the meridian, then the hour angle of the celestial body when on the vertical wire will be zero; if this is not so, it becomes necessary to determine the hour angle of the star when in that line, in order that we may reduce the observation to that instant of time when the hour angle is zero; in other words, to determine the hour angle of the star O which is Z P O, in time; this applied to the clock time of the observed transit will give the time of the meridian transit. This angle is the *reduction to the meridian* and evidently depends upon the three quantities a, b, and c, above mentioned. To determine the value of this angle, let L represent the latitude of the observer; b, as we have already seen, is the altitude of the point A; let n represent the declination of A, and m, the complement of its hour angle. Then in the triangle Z P A we have—

$$\mathrm{ZPA} = 90^\circ - m \qquad \mathrm{ZA} = 90^\circ - b$$
$$\mathrm{PA} = 90^\circ - n \qquad \mathrm{PZA} = 90^\circ + a$$
$$\mathrm{PZ} = 90^\circ - \mathrm{L}$$

and, substituting these values in equations (6) and (4) of Spherical Trigonometry, we have—

$$\cos n \sin m = \sin b \cos \mathrm{L} + \cos b \sin a \sin \mathrm{L} \qquad (1)$$
$$\sin n = \sin b \sin \mathrm{L} - \cos b \sin a \cos \mathrm{L} \qquad (2)$$

these determine m and n, when a and b are given. Again, in the triangle APO let τ represent the hour angle of the star O, when on the middle vertical wire, that is, the angle ZPO; let d represent the declination of the star, and c, as before, the collimation constant. We shall then have—

$$\mathrm{A\,P\,O} = \mathrm{A\,P\,Z} + \mathrm{Z\,P\,O} = (90^\circ - m) + \tau = 90^\circ + (\tau - m)$$
$$\mathrm{A\,P} = 90^\circ - n \qquad \mathrm{A\,O} = 90^\circ + c$$
$$\mathrm{P\,O} = 90^\circ - d.$$

which give the following equation:—

$$-\sin c = \sin n \sin d - \cos n \cos d \sin (\tau - m) \qquad (3)$$

or

$$\sin (\tau - m) = \tan n \tan d + \sin c \sec n \sec d \qquad (4)$$

determining $\tau - m$ and consequently τ when c, b, and a, are given. These are the general formulæ for the reduction to the meridian.

153. Having, by the method already described, placed the instrument nearly in the plane of the meridian, the angles a, b, and c are reduced to very small angles and it follows that m, n, and τ will also be very small. Therefore we may substitute for the sines of these small angles the angles themselves, and, for their cosines, unity ; then

(1) becomes $m = b \cos \mathrm{L} + a \sin \mathrm{L}$ (5)

(2) becomes $n = b \sin \mathrm{L} - a \cos \mathrm{L}$ (6)

(4) becomes $\tau = m + n \tan d + c \sec d$ (7)

Equation (7) is known as Bessel's formula.

Substituting the values of m and n in (7) factoring and reducing we have

$$\tau = b \frac{\cos(\mathrm{L} - a)}{\cos d} + a \frac{\sin(\mathrm{L} - d)}{\cos d} + \frac{c}{\cos d} \qquad (8)$$

known as Mayer's formula for computing the reduction to the meridian, in other words, the correction to be applied to the observed sidereal clock time of the transit of a celestial body over the middle thread, to obtain the *clock* time of its transit over the meridian. Let T be the *observed* clock time of transit over the middle thread, then $\mathrm{T} + \tau$ will be the clock time of transit over the meridian. The exact sidereal time of the transit of a star is known from its right ascension. Hence, comparing our clock time of transit over the meridian with the right ascension of the star taken from the Star Catalogue we are enabled to determine the correction for the clock. Let $\Delta \mathrm{T}$ represent that correction, and A the right ascension of the star; then

$$\mathrm{A} - (\mathrm{T} + \tau) = \Delta \mathrm{T} \qquad (9)$$

substituting the value of τ.

$$\Delta \mathrm{T} = \mathrm{A} - \left(\mathrm{T} + b \frac{\cos(\mathrm{L} - d)}{\cos d} + \frac{c}{\cos d} + a \frac{\sin(\mathrm{L} - d)}{\cos d} \right) . \qquad (10)$$

$$\mathrm{A} = \mathrm{T} + \Delta \mathrm{T} + b \frac{\cos(\mathrm{L} - d)}{\cos d} + \frac{c}{\cos d} + a \frac{\sin(\mathrm{L} - d)}{\cos d} \qquad (11)$$

by (10) the correction of the clock can be found from the observed transits of known stars, and by (11) having the clock correction the right ascension of stars not catalogued can be determined.

154. The Coast Survey Office has published "Tables of Factors

for Reduction of Transit Observations," by means of which the factors

$\frac{\sin (L - d)}{\cos d}$, $\frac{\cos (L - d)}{\cos d}$, and $\frac{1}{\cos d}$ may be taken directly from

the table using $(L - d)$ and d as arguments.

In very exact observatory work a small correction caused by diurnal aberration should be applied to the apparent right ascension; the formula for this is $-\frac{0^{sec}.021 \cos L}{\cos d}$. As will be seen, however, its value would scarcely be appreciable.

155. The above values of τ are to be applied to the upper culminations of celestial bodies. For lower culminations it is necessary to substitute for A, 12 hrs. + A, and for d, $180° - d$, since in the formulæ d has represented the distance of the body from the equator reckoned toward the zenith, and since the time of lower culmination differs by 12 hours sidereal time from that of the upper culmination of the same body.

EQUATORIAL INTERVALS.

156. Heretofore we have supposed that the observations were made upon the middle vertical thread; this method of observing, however, is rarely adopted, because a single observation is more liable to error than the mean of several. The diaphragms of all astronomical instruments are constructed with an odd number of threads (in this particular instrument nine), and these are placed parallel and as nearly equidistant as possible. If they were exactly equidistant the mean of the observed times of transit over all could be taken as the time of transit over the middle thread, but this is very rarely the case even in the largest instruments; hence it is necessary to determine the distances between the threads so that we may reduce our observations either to the middle thread or to the *mean thread*, the latter being the one generally used with the portable instrument.

The *mean thread* may be defined as an imaginary thread so placed in the field that the time of transit over it will be the same as the mean of the times of transit over all the threads. All observations are reduced to this imaginary thread. The constants a, b, and c, and the intervals of the several threads are referred to it. As it often happens that it is only possible to observe transits over a portion of the threads, it is evidently necessary, in order to deduce the time of transit over the mean thread, that the distance of each from the mean thread be known. The determination of these distances is usually the first thing done after the instrument has been placed approximately in the meridian and adjusted to sidereal focus. The distance required is evidently the angle at the eye piece subtended by the interval between the imaginary thread and the thread given, and this might be obtained by observing an equatorial star, that is, one moving exactly in a great circle. Note the times of transit over all the wires, the mean of these will be the time of transit over the mean wire; and the difference between this and each time noted will be the equatorial interval of the noted thread, called equatorial interval when expressed in time. On account of the difficulty of obtaining conviently an exact equatorial star, the practical method is to observe a slow moving star and reduce its intervals in time to the great circle intervals by the following method.

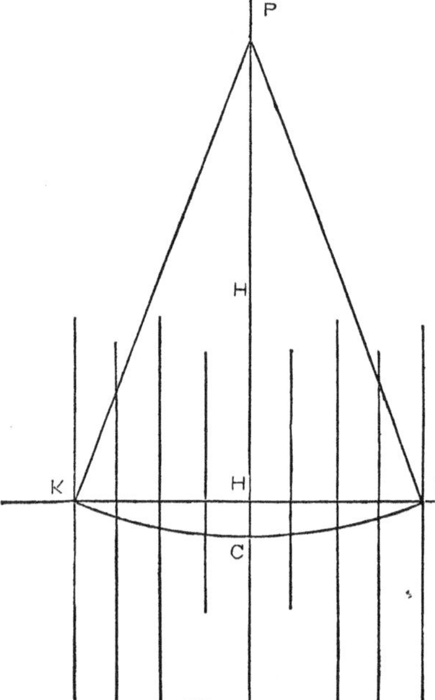

Fig. 22.

In the figure let us suppose the mean thread H to correspond exactly with the meridian and all threads to be parallel to it.

Let F C K represent the path of a close polar star in the field and P the pole. Now if the times when the star is on each successive thread be noted and the mean taken, we shall have the time when the star is on the mean thread, and the difference, between this last time and the time noted, when the star was at F will evidently be the hour angle of the star when at F the angle F P H. It is required to find the great circle distance F H which is done by solving the right angled triangle F P H in which F P H is known and P F is the polar distance of the star. Let i_1 represent the required equatorial interval of thread *one*, and I_1 the noted interval in time between the mean thread and thread one, that is, the angle F P H; then

$$\sin i_1 = \sin I_1 \cos d;$$

d representing the declination of the star. This will be rigidly true only when the mean thread corresponds with the meridian and when the other threads are truly parallel to it. Our adjustments, however, are approximate and, in practice, having made them as close as possible, the above formula will be sufficiently exact for use in determining the equatorial intervals required. It shows the importance of making the first adjustments with extreme care.

157. In cases where d is less than 80° I_1 and i_1 will be very sma l angles, and we may substitute the angles for their sines; or

$i_1 = I_1 \cos d$ transposing
$I_1 = i_1 \sec d$ by means of which

having determined the equatorial interval of any thread, the time it will take a star, whose declination is known, to travel from that thread to the mean thread, can be found.

The work will be easily understood from the following:—Let $t_1\ t_2\ t_3; \ldots . t_9$ be the observed times of transit over the successive threads; $i_1\ i_2\ i_3 \ldots . . i_9$, the equatorial intervals of each successive thread from the mean thread; d = the declination, and

$t = \dfrac{t_1 + t_2 + t_3 \ldots \ldots t_9}{9}$ equal the time of transit over the mean

thread, we shall then have, $i_1 = (t - t_1) \cos d,$

$$i_2 = (t - t_2) \cos d,$$
&c., &c.. &c.
$$i_9 = (t - t_9) \cos d,$$

and the intervals will have the proper signs for use to find the correction. $I_n = i_n \sec d$ to be used in broken transits, Clamp west.

Should the rate of the clock be found to be great, it will be necessary to correct our intervals $(t - t_1)$ &c., for its change using

$$1 + r_m \ (t - t_1) \ \&c.,$$

r_m being the gain or loss per minute, and $(t - t_1)$, &c., expressed in minutes and decimals: r_m, positive, when clock is losing; negative, when gaining.

The threads are usually numbered I. II. III. IV. V. VI., VII. VIII. IX. in order, commencing with the one nearest the clamp end of the axis, so that (since the telescope apparently reverses the motion of the star in the field) with the Clamp west, the star at its upper culmination will cross the threads in regular succession I. II. . . . . IX., while with Clamp east it will cross IX. VIII. . . . . I. Equatorial intervals, once determined, are marked for the threads I. II. III. &c., IX. ; hence it is necessary to be careful with their signs, which will change in the corrections upon shifting the position of the axis. The times of transit are noted at the instant when each wire bisects the star as it crosses the field.

158. Having determined the equatorial intervals i_1 i_2, &c., I_n, the interval in time between the mean thread and a star on any other thread can evidently be found by the formula $I_n = i_n \sec d$. Whence, noting the transit over any thread of a star whose declination is given, the time of its transit over the mean thread can be determined. In case of a "broken transit," where some of the threads are missed, the time t required is equal to the mean of the observed times, minus the sum of the equatorial intervals of missed threads divided by number of observed threads multiplied by sec d, for, as already seen,

$t = \dfrac{t_1 + t_2 + t_3 + \dots t_9}{9}$. Suppose t_2 was not noted, we have $i_2 = (t - t_2) \cos d$, whence $t_2 = t - i_2 \sec d$; substituting this value of t_2 we have $t = \dfrac{t_1 + t_3 + \dots t_9 + t - i_2 \sec d}{9}$

transposing $t = \dfrac{t_1 + t_3 + \dots t_9}{8} - \dfrac{i_2 \sec d.}{8}$

159. If the transit over one, or a few threads only, be observed, we may use $t =$ mean of observed times, plus the sum of equatorial intervals of observed threads divided by number of observed threads multiplied by sec d. Since from preceding

$$\left.\begin{array}{l} i_1 = (t - t_1) \cos d, \text{ or } t = i_1 \sec d + t_1 \\ i_2 = (t - t_2) \cos d, \text{ or } t = i_2 \sec d + t_2 \end{array}\right\}$$ and the mean of these would

give $t = \dfrac{t_1 + t_2}{2} + \dfrac{(i_1 + i_2)}{2} \sec d$.

Remembering always that as determined, the equatorial intervals have their signs as correction, and these signs must be carefully used.

LEVEL CONSTANT.

160. In the formula for the reduction to the meridian

$$\tau = a \frac{\sin (L - d)}{\cos d} + b \frac{\cos (L - d)}{\cos d} + \frac{c}{\cos d}$$

a numerical value in time to be applied to the observed time of transit over mean thread, as a correction, can be found, if the value of a, b, and c are known. The first determined is b, the level constant, and the method is as follows ;—Place the striding level in its position on the axis, and read the divisions at each end of the bubble, mark the reading of W. end W, and of E. end E ; it is evident then that the reading of the centre of the bubble would be the mean of the two end readings, or $\frac{W - E}{2}$ Reverse the level and again take the end readings of the bubble, marking W. end W′ and E. end E′ as before, the expression for the reading of the centre of the bubble will be $\frac{W' - E'}{2}$. If the axis be level, the two readings of the centre of the bubble will be equal with opposite signs, but if not, the *mean* of these two readings will express the number of level divisions, showing its deviation from the level, or $\frac{1}{2}\ \frac{(W - E)}{2} + \frac{(W' - E')}{2}$ = deviation in terms of level divisions. The value of the level division may be determined by the level trier, or by securing the level to any well graduated alidade, and noting the readings of the centre of the bubble as follows : Having secured the level to the graduated circle of the Meridian Circle at the Observatory, bring the bubble near one end of the tube, and note its end readings, the mean of them will be the central reading ; by means of micrometers attached, note the reading of the circle, then, by means of clamp screw of circle, bring level bubble near other end of tube, and again find its central reading : the difference between the two central readings will be the number of level divisions corresponding to the angle through which the circle has been turned, determined accurately by the micrometer readings in the two positions of circle : the value of one division will be this angle divided by the whole number of divisions noted. The expression for b will then be $\frac{1}{2}\left(\frac{(W - E)}{2} + \frac{(W' - E')}{2}\right) d$, letting d represent the value of level division, transposing, b in arc $= \frac{(W + W') - (E + E')}{4}\ d$; to reduce to time divide by 15 $\therefore b = \frac{(W + W') - (E + E')}{60}\ d.$: this constant is marked W. or E, according to the name of the greater, and signifies that the corresponding end of the axis is too high ; for use in our formula it is positive when W. end is too high, as is mathematically shown by its expression.

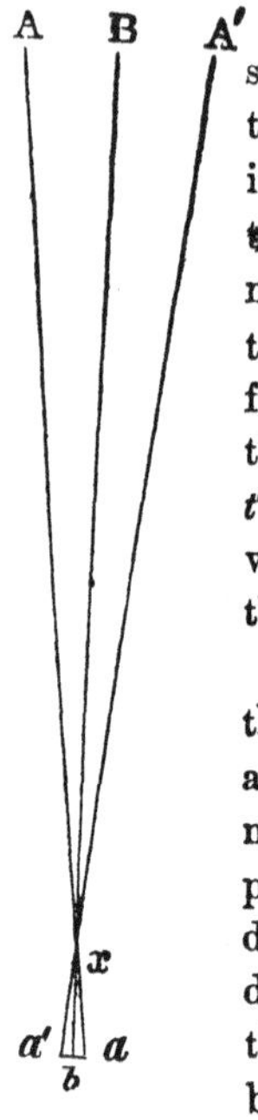

A B A′

x

a′ a
b

161. *To find c, the Collimation Constant,* observe a slow moving star at upper culmination. Note the transit of the star over the first half of the threads, including or excluding the middle thread ; then reverse the axis, and note the transit over the same threads now in reverse order. By means of the equatorial intervals, find the time of transit over the mean thread from each of these sets of observations ; then let t be the time of transit over *mean thread* before reversal, and t' the time of transit over *mean thread* after reversal ; we shall then have $c = \frac{1}{2}(t' - t) \cos d$, its value for that position of the axis before reversal.

In the figure ; suppose that the mean thread occupies the position a out of the axis of collimation b B ; c, the angle required is $a\,x\,b$ or A x B ; when the instrument is reversed the mean thread will occupy the position a', and $a\ x\ a' =$ A x A′ will evidently be double the required angle. By observation we have deduced the times of transit of the same star over the mean thread in its two positions, and the difference between these times is an expression for the value of the angle A x A′, which, reduced to great circle measure and divided by two, will be the angular value in time of c, as above given.

Since the instrument may be used in two positions of the rotation axis, it is necessary to distinguish them, and this is done by noting the position of the clamp end of the axis, marking all observations Clamp east or Clamp west. The collimation constant c will evidently depend for its sign of application upon the position of the axis ; for, having been determined with Clamp west, say, positive in position a the star crossing the mean thread earlier than it crosses the continuation of the axis of collimation, when reversed, it will cross that thread later, hence the sign of c Clamp east would be negative, the numerical value remaining the same. For lower culminations the formula becames $\frac{1}{2}(t - t') \cos d$, giving the proper sign of c for the position of the axis before reversal. It may be necessary in computing t and t' to apply a correction for the rate of the Chronometer, as has already been explained in connection with equatorial intervals.

To find a, the Azimuth Constant.

162. Note the times of transits of two stars differing as much as possible in declination and having nearly the same Right Ascensions, so that the interval between the observations shall be short.

Let A = Right Ascension.
d = declination.
and T = observed clock time of
transit over the mean thread of the first star and A′, d' and T′, the same respectively of the second star. From (11) we have

$$A = T + \Delta T + b\,\frac{\cos(L-d)}{\cos d} + \frac{c}{\cos d} + a\,\frac{\sin(L-d)}{\cos d} \text{ and}$$
$$A' = T' + \Delta T' + b\,\frac{\cos(L-d')}{\cos d'} + \frac{c}{\cos d'} + a\,\frac{\sin(L-d')}{\cos d'}$$

If the clock has no rate, Δ T, the correction at time of first observation will be equal to Δ T′ the correction at second observation, and the two equations will eliminate the unknown correction
Δ T = Δ T′. If the rate be known (generally the case)
$\Delta T' = \Delta T + r_m (T' - T)$, in which r_m is the rate per minute, and $(T' - T)$ is expressed in minutes and fractional parts. Substituting for Δ T′ in above, we have

$$A' = T' + r_m (T' - T) + b\,\frac{\cos(L-d')}{\cos d'} + \frac{c}{\cos d'} + \Delta T + a\,\frac{\sin(L-d')}{\cos d'}$$
$$A = T + b\,\frac{\cos(L-d)}{\cos d} + \frac{c}{\cos d} + \Delta T + a\,\frac{\sin(L-d)}{\cos d},$$

in which the unknown quantities are Δ T and a. Since r_m, b, c, L, d and d' are known, we will substitute

$$t' = T' + r_m (T' - T) + b\,\frac{\cos(L-d')}{\cos d'} + \frac{c}{\cos d'} \text{ and}$$
$$t = T + b\,\frac{\cos(L-d)}{\cos d} + \frac{c}{\cos d};$$

in other words, t' and t are the observed times of transit corrected for level and collimation errors, and also the rate of the clock. Our formulæ will then be:

$$A' = t' + \Delta T + a\,\frac{\sin(L-d')}{\cos d'} \text{ and}$$
$$A = t + \Delta T + a\,\frac{\sin(L-d)}{\cos d}$$

Subtracting, $A' - A = t' - t + a\left(\frac{\sin(L-d')}{\cos d'} - \frac{\sin(L-d)}{\cos d}\right)$

Transposing, $a = \dfrac{(A' - A) - (t' - t)}{\dfrac{\sin (L - d')}{\cos d'} - \dfrac{\sin (L - d)}{\cos d}}$, the formula for determining the azimuth constant, in which the two terms in the denominator may be taken from the "Coast Survey Tables for Reduction of Transit Observations" directly.

By Chauvenet's Plane Trig., (172), the denominator of the fraction above $\dfrac{\sin (L - d')}{\cos d'} - \dfrac{\sin (L - d)}{\cos d}$ becomes $\dfrac{\cos L \sin (d - d')}{\cos d \cos d'}$, and the expression is $a = [(A' - A) - (t' - t)] \dfrac{\cos d \cos d'}{\cos L \sin (d - d')}$ which is the formula used for adjusting the instrument in the first part of the work; approximate only, since we used uncorrected times of transit. Again,

$\dfrac{\cos d \cos d'}{\cos L \sin (d - d')} = \dfrac{1}{\cos L (\tan d - \tan d')}$, which substituted give,

$a = \dfrac{(A' - A) - (t' - t)}{\cos L (\tan d - \tan d')}$, a more convenient form in case no "Tables" are at hand.

163. Now having the three constants a, b, and c, the value of τ, the reduction to the meridian, may be found, from (8), and, by means of (9) Δ T, the correction for the clock or chronometer may be accurately determined, and hence the exact determination of absolute time is made.

164. The right ascensions and declinations of stars are taken from the American Ephemeris but, as the number there is rather limited, special catalogues for use in the Northern Hemisphere have been prepared for the use of the Coast Survey. The British Almanac may be used if at hand, but for particular work it would be well to have stars whose positions have been accurately fixed at the Observatory at the latest possible moment.

165. In making a set of observations for time, the series generally commences with those stars observed for making the instrumental corrections, then the accurately determined time stars, and afterward another set for instrumental corrections in order to determine whether or not there has been any change in them. The axis of the instrument

should be reversed, so that one-half the observations should be with Clamp east and the other half with Clamp west, and as many stars are taken as possible, the mean of all the observations being used. The same observer should make all the observations, and if his *personal equation* is known the best results are to be expected.

166. For further and more detailed information in regard to this instrument, the reader is referred to Chauvenet's Astronomy, Vol. II. and for practical information in its use, to the report for 1866 and the various pamphlets issued at the Coast Survey Office.

The methods given are by no means the only ones, either for the practical or the theoretical adjustments, and some of the more refined corrections have been omitted. The error for the inequality of the pivots has not been referred to, as it is caused by poor workmanship, generally reduced to a minimum in those instruments accepted by the Government. Whatever error there may be should be determined at the Observatory, and this, with the value of the level division, should be furnished with the record of the instrument.

CHAPTER XIII.

THE ZENITH TELESCOPE—ADJUSTMENTS—TALCOTT'S METHOD FOR LATITUDE—CONSTANTS OF THE INSTRUMENT—METHOD OF OBSERVATION—REDUCTION TO THE MERIDIAN—SELECTION OF STARS.

ZENITH TELESCOPE. (See PLATE.)

167. In this instrument the telescope tube is attached to one end of the horizontal axis, counterpoised at the other end by a weight so connected as to prevent flexure of the axis and to equalize the pressure upon the two Ys. The two Ys are supported by a single vertical standard, which contains the vertical axis of the instrument, and this standard revolves about the vertical axis, carrying a clamp and vernier by means of which it may be set to any reading of the horizontal circle at its base. The frame is fitted with three foot screws at its base for the purpose of adjusting the verticality of the vertical axis, and the striding level is applied to the horizontal axis as in the case of the transit instrument, so that when horizontal the line of collimation of the telescope will move in a vertical plane. Lamp attachments are also fitted so that the field may be illuminated for the observation of faintly illumined bodies. This instrument is especially used for the determination of small differences in zenith distances, particularly in determining the latitude by Talcott's method. As furnished for use, it generally consists of a telescope of 45 inches focal length, with aperture of $3\frac{1}{2}$ inches and magnifying power from 60 to 120 so that stars of the 6th and 7th magnitude may be conveniently observed. The horizontal axis is about 7 inches in length, the vertical standard about 24 inches, and the diameter of the horizontal or azimuth circle is 12 inches, graduated to read to half minutes

and less with the vernier; two movable stops can be applied to this circle defining on the instrument the position of the plane of the meridian without interfering with the motion while reversing. Attached to the telescope, concentric with the horizontal axis, is a graduated (vertical) circle on which zenith distances are read off to within 30″ by means of the arm and vernier attached to a very delicate spirit level (value of one division about $\frac{3}{4}$ of a second). The telescope can be set to any inclination and clamped, while for accurate pointing the bubble is brought into the middle by means of a fine motion screw as in the case of the transit instrument.

In the eye tube is the micrometer for the accurate measurement of zenith distances. The screw carries one or more horizontal threads and its head is divided in 100 parts, of which tenths may be estimated, the whole number of turns made by the screw is indicated by a rack shown on the side of the field of view. The value of one complete revolution is generally less than 50″ so that observations can be read exactly to 0″.5 and by estimation within less than 0″.05. There are usually added several fixed vertical threads, so that the instrument can be used as a transit when required. For convenience of observation the telescope is fitted with the diagonal eye piece.

ADJUSTMENTS.

168. When setting up the instrument it will be found convenient to place two of the foot screws in the E. and W. line, proceed then to make the vertical axis truly vertical by means of the striding level which should not change when the instrument is made to describe a complete revolution in azimuth, this test to be fulfilled by means of the foot screws, assisted by a small circular level also attached to the frame; next test the horizontality of the transit axis by reversing the striding level as in the case of the transit instrument; adjust the eye piece to sidereal focus, and see that the micrometer thread is horizontal, this last is proved by an equatorial star running along the thread, or by the same appearance of a polar star, when the instrument is turned in azimuth. The adjustment of the line of collimation is effected by means of a distant, well defined, terrestrial object; turning the instrument so that the readings of the azimuth circle shall be just 180° apart and allowing for the parallax, since the line of collimation is not in the plane of the vertical axis. The amount of this parallax can easily be computed in a given case; for if $d =$ the distance of the centre of the telescope

from the vertical axis, D = the distance of the object, and p = the parallax, we have

$$p = \frac{d}{\text{D} \sin 1''}$$

but, as the horizontal circle is not designed for very accurate measures, it will not usually be worth while to use this method further than to make a first adjustment. A perfect adjustment, however, may be made by two collimating telescopes. The reading on the horizontal circle, of the plane of the meridian, is ascertained by means of the known chronometer time of the culmination of a slow-moving star, which is bisected at that time by the middle thread and the corresponding reading of the circle noted; stops are then applied to indicate the meridional position with the telescope pointing N. or S. of the zenith.

TALCOTT'S METHOD FOR DETERMINING LATITUDE.

169. From general formulæ, we have for body south of zenith

$$\text{L} = z + d$$

while for body north of zenith

$$\text{L} = d' - z'$$

(L representing the lat., d and d' the declinations, and z and z' the zenith distances of the two bodies respectively.) Combining the two equations we have

$$2\,\text{L} = (d + d') + (z - z')$$

$$\text{or L} = \tfrac{1}{2}\,(d + d') + \tfrac{1}{2}\,(z - z'),$$

in which it is seen that, if the declinations are correctly known and the *difference* of the zenith distances found, the determination of the lat. is correct.

The difference of the zenith distances is measured by the micrometer in the zenith telescope, and the method of observation is as follows :—

Two stars, technically called a pair, are so selected that the difference of their zenith distances shall be less than the breadth of the field of the telescope. The instrument, adjusted to the plane of the meridian, is set at an altitude corresponding to the mean of the two zenith distances and pointed to that star which transits first, the bubble of the level made to play nearly in the middle of the tube. When the star enters the field, take up the beat of the chronometer and bisect the star with the micrometer thread; keep it so bisected until the Chro. time of culmination, when note the reading of the micrometer thread, and also the reading of the level ; revolve the telescope through 180° in azimuth (marked by the stops), and observe the second star in the same manner. The difference of the micrometer readings, corrected for difference of level and refraction, will evidently be the difference of the zenith distances required ; and one-half this difference applied to the half sum of the declinations will give us the latitude of the station. It remains for us to determine the reading of the micrometer, and of the level, and the correction for refraction.

170. Let us assume the zero reading of the micrometer at that part of the apparent field farthest from the zenith ; and let Z_0 represent the apparent zenith distance of this 0 reading, when the level reading is 0. Now point the telescope to the first of a pair of stars, say the one S. of the zenith, and bisect it with the movable thread ; let m represent the number of turns given to bring the thread to the star ; and let R represent the value of a single turn. If the level reading still remains 0, the expression for the apparent zenith distance of the star will evidently be $Z_0 - mR$. If the level has changed, however, let l represent the number of divisions through which it has changed, and D the value of a division ; and suppose the north end is too high, then Z, the true zenith distance of the first star, will be equal to $Z_0 = mR + l\,D + r$, letting r represent the refraction which tends to decrease the (true) zenith distance.

Turn the instrument through 180° in azimuth and point to the N. star of the pair, bisect as before, and let m' be the number of micrometer turns and l' the number of level divisions noted in this case, then the true zenith distance of the second star, $Z' = Z_0 - m'R - l'D + r'$

Subtracting we have

$z - z' = (m' - m)\,R + (l' \times l)\,D + (r - r')$ and

$\frac{1}{2}(z - z') = \frac{1}{2}(m' - m)\,R \mp \frac{1}{2}(l' + l)\,D + \frac{1}{2}(r - r')$.

These terms for use in determining the half difference in zenith distance will be considered in detail.

REFRACTION.

171. A principle of refraction is, that when zenith distances are small, it varies as the tangent of the zenith distance ; and, since observations are generally restricted to zenith distances of less than 25° in this method, the formulæ which follow may be employed without sensible error.

$$r = a \tan z$$
$$r' = a \tan z' \qquad \text{whence}$$
$$r - r' = a (\tan z - \tan z')$$
$$\tan z - \tan z = \frac{\sin (z - z')}{\cos z \cos z'}$$
$$\therefore r - r' = a \frac{\sin (z - z')}{\cos z \cos z'}$$

Since z and z' are nearly equal, we may substitute $(z - z') \sin 1'$ for $\sin (z - z')$ and $\cos^2 z$ for $\cos z \cos z'$, when the formula becomes

$$(r - r') = (z - z') \frac{a \sin 1'}{\cos^2 z}$$

a is the coefficient used in Bessel's formula for refraction, which, although strictly a variable, is in this connection considered constant, since the variation in z is small, and any variation in Bar. or Ther. is neglected; the value assumed for a in coast survey work is 57.″7. For use in our formula we have $\frac{1}{2} (r - r') = \frac{1}{2} (z - z') \frac{a \sin 1.}{\cos^2 z}$ the last term of which may be tabulated (see Chauvenet's Astro., Vol. II), or as has been done by the Coast Survey Office; $\frac{1}{2} (r - r')$ can be taken from the table directly, using $\frac{1}{2}$ the difference of zenith distances and the zenith distance as arguments. Since the correction is very small, it will be sufficiently accurate to use $\frac{1}{2} (m' - m)$ R instead of $\frac{1}{2} (z - z')$. The sign of the correction then is the same as that of the correction for the micrometer.

LEVEL.

172. For reading the number of divisions of the level we have for the S. star $l = \frac{n - s}{2}$ and for the N. Star $l' = \frac{n' - s'}{2}$ in which n is the reading of the N. end of the bubble and s of the S. end; then $\frac{1}{2} (l + l') = \frac{(n + n') - (s' + s)}{4}$, which will give us the proper sign for the application of this correction.

MICROMETER.

173. $\frac{1}{2} (m' - m)$ would evidently be read by taking the number of teeth in the rack passed over by the thread and adding the reading of the micrometer head thereto.

VALUES OF D AND R.

174. If now we determine the values of R and D, all the terms in our equation become known. If the value of the level division D be accurately determined by the level trier or by the method already explained in connection with the transit, we can from that determine the value of R, the revolution of the micrometer screw, as follows: Direct the telescope to some distant well-defined point (better the cross-threads of a collimating telescope) and bisect it with the micrometer thread. Clamp the level, and by means of the small tangent screw in connection with it, bring the bubble near one end of the tube, and note the reading. Now, by means of the tangent screw, moving the telescope and level so attached, bring the bubble to the other end of the tube and note its reading; it is evident that by so doing we have moved our line of sight, (over the undisturbed micrometer thread), through an angle whose value is expressed by $(l-l')$ D, $(l - l')$ being the number of divisions through which the bubble has passed. Now, move the micrometer thread back until it exactly bisects the same distant point; by so doing we have moved it through this angular distance $(l - l')$ D; noting the number of turns given the micrometer screw $(m-m')$ and dividing, we have $R = \frac{(l-l')\ D}{(m - m')}$ Repeating this operation with the micrometer thread in different parts of the field, and taking the mean, a very accurate result will be obtained, and so both values will be known.

175. The method employed in the coast survey is to first determine the value of the level division in terms of R. and this is done in precisely the same method as that just described, the resulting equation being

$$D = \left\{ \frac{m-m'}{l-l'} \right\} R. \quad \text{Let } d = \frac{m-m'}{l-l'}\ ;\ \text{then } D = d\,R.$$

TO DETERMINE THE VALUE OF A REVOLUTION OF THE MICROMETER SCREW.

176. This may be done by turning the micrometer through an angle of 90° on the telescope, and noting the times of passage of a circumpolar star at culmination over the thread, placed successively before the star for each turn of the screw, and treating the intervals as Equatorial intervals. (See "Transit Instrument.")

177. It is evident that the same principle will hold good if the observa-

tions be made upon a polar star near the time of its elongation; in that case, the *vertical* thread of the instrument, if directed to the star at the instant of elongation, will be tangent to the diurnal circle, and the right angled spherical triangle would be solved as follows:

Let s. s'. s'' be the apparent path of a star in the field of the instrument, a b the vertical thread, and suppose m m' m'' to represent the successive positions of the micrometer thread placed in advance of the star. To find the distance $s\ s''' = i$, in great circle units, we have to solve the right angled triangle P s''' s. P s being the polar distance, $sps' =$ I, the noted interval of time between the observations of the star on m and m', or the hour angle of the star from elongation,

$$\sin i = \sin \mathrm{I} \cos d.$$

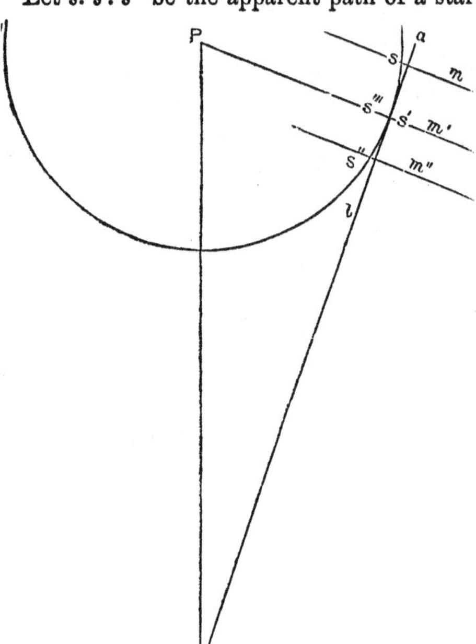

Fig. 24.

which is precisely the same form as that used in determining Equatorial intervals.

Since the instrument can be easily turned in azimuth, it will be seen that the last method is to be preferred, as it involves no change in the micrometer or in the focus of the telescope.

METHOD OF OBSERVATION.

178. Compute the time of the star's elongation, its zenith distance and azimuth at that time, from the right angled spherical triangle. P Z. s'.

$$\cos t = \cot d \tan \mathrm{L}.$$
$$\cos z = \operatorname{cosec} d \sin \mathrm{L}.$$
$$\sin \mathrm{Az.} = \cos d \sec \mathrm{L}.$$

By means of these formulæ point the telescope to the star twenty or thirty minutes before the Chro. time of elongation, and bisect it with the micrometer thread; note the time of bisection, the reading of the micrometer and the level reading. Move the thread in advance of the star by one or more revolutions of the screw and again note time, level, and micrometer readings. Repeat the observations as many times as possible while the star is in the field. Let t_0 represent the Chro. time of elongàtion, t t_1 t_2 &c., the noted Chro. times of bisection and m m_1 m_2 &c., the corresponding micrometer readings. Suppose m_0 to be the micrometer reading of the star at time of elongation, and let i i_1 i_2 be the required angular distances respectively, then:

$$\sin i = \sin (t_0 - t) \cos d$$

$$\sin i_1 = \sin (t_0 - t_1) \cos d.$$

If the level has remained constant, we have

$$(m_0 - m) \text{ R} = i$$

since $(m_0 - m)$ represents the number of turns given the micrometer screw to move the thread through the angular distance i, also

$$(m_0 - m_1) \text{ R} = i_1$$

subtracting

$$(m_1 - m) \text{ R} = i - i_1$$

and

$$\text{R} = \frac{i - i_1}{m_1 - m}.$$

Should the level have changed its reading it is evident that a portion of our angular distance $(i - i_1)$ is due to the change of direction of the axis of the telescope. To make the correction for this change, let l represent level reading when micrometer reading is m and l_1, level reading when micrometer reading is m_1, then $(l_1 - l)$ D is the change required, the angular value of D is unknown, but we have found

$$\text{D} = d \text{ R},$$

hence the correction is $(l_1 - l)\, d$ R. to be applied to $i - i_1$, *i. e.*

$$(m_1 - m) \text{ R} = (i - i_1) \pm (l_1 - l)\, d \text{ R, or}$$

$$\text{R} = \frac{i - i_1}{(m_1 - m) \mp (l_1 - l)\, d}$$

179. Deducing values of R for each of the observations taken—the mean will be the very accurate value required. It is to be afterward corrected for refraction by the following method.

Find the change in refraction for a change of 1′ in zenith distance, at the computed zenith distance, from the refraction tables. Let dr represent that change, then R dr will be the correction to be subtracted from our value of R, just found.

Now having R, D the value of the level division is found from

$$\text{D} = d \text{ R}.$$

In case of there being more than one movable thread in the micrometer, the angular distance between them may be measured by noting the number of revolutions given the screw to bring one thread to the position occupied by the other; to be done by bisecting a distant point by each wire successively, carefully noting any change in the level reading, and applying the correction as before.

REDUCTION TO THE MERIDIAN.

180. Since it may not always be possible to obtain an exact meridian observation, a single observation off the meridian may be taken in two ways, and by means of the "*reduction to the meridian,*" reduced.

One of the methods is to follow the star, turning the instrument in azimuth, and measure the exact zenith distance—noting the time. The reduction then is the same as for circum-meridian altitudes. (See Coffin's Navigation, p. 120.)

The other method is to bisect the star off the vertical wire noting the time. The "reduction" in this case will be understood from the figure.

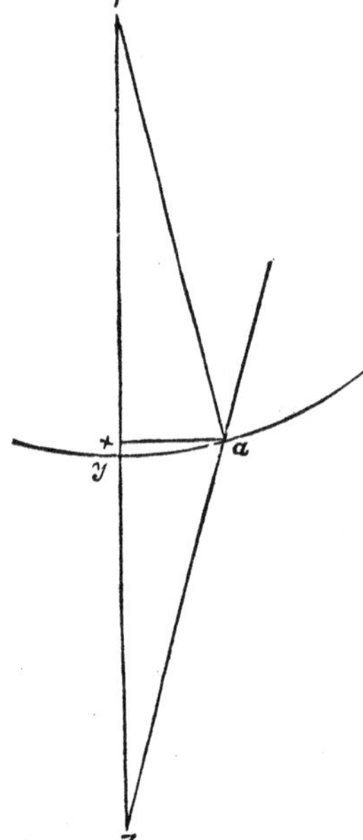

Fig. 25.

The vertical thread being in the plane of the meridian, the micrometer thread xa at right angles to it bisecting the star, evidently measures the distance Z x, differing from the meridian zenith distance by xy, the "reduction to the meridian." To find this value solve the right angled triangle $x\text{P}a$, for Px. Py = Pa =polar distance of star x P a = hour angle of star, then $\tan \text{P}\,x = \cos t \cot d$.

$$xy = p - \text{P}x$$

This value of the reduction will evidently be negative or subtractive from the observed zenith distance when the star passes between the pole and the zenith, and positive in all other cases. The angle t should always be small.

181. Referring to the figure, it will be seen that in the first method moving our instrument in azimuth gives the vertical

thread a direction *za*, and the micrometer thread being perpendicular to it, the distance measured is the apparent zenith distance *za*. The reduction in this case may be taken from Bowditch's Tables.

Other methods are given in Chauvenet's Astronomy and in the Coast Survey Publications.

SELECTION OF STARS.

182. It will in general be necessary to obtain star catalogues especially prepared for this work, as the ordinary ephemerides are not sufficiently large to furnish a sufficient number of *pairs*. Such catalogues have been prepared for use in Northern Latitudes by especial observations. It is evident that a mistake made in the exact position of a star employed will effect the latitude obtained by this method, therefore it is recommended to select those stars only which have a very satisfactory authority. Those taken from the catalogues of the Washington Observatory, and of the Greenwich Observatory, checked with each other, will furnish one or two dozen pairs for each night. The stars selected should differ as little as possible in zenith distances, never more than 50′ with the most improved and latest instruments. The interval of time between the culminations of a pair should be as short as possible to guard against any possible change of the instrument, and yet sufficiently long to permit the deliberate reading of the micrometer and the level and allow the instrument to be turned in azimuth with a small margin for taking up the star and bisecting it.

183. The *mean* places of stars are given in the Catalogues. To determine the *apparent* places, which are the ones to be used, the student is referred to Chauvenet's Astronomy, Vol. I., art. 77. Personal errors and probable errors will be found thoroughly discussed in Chauvenet's Astronomy, Vol. II., and in the Coast Survey Pamphlets, to which the student, who intends making himself proficient as an observer, is referred.

Plate I.

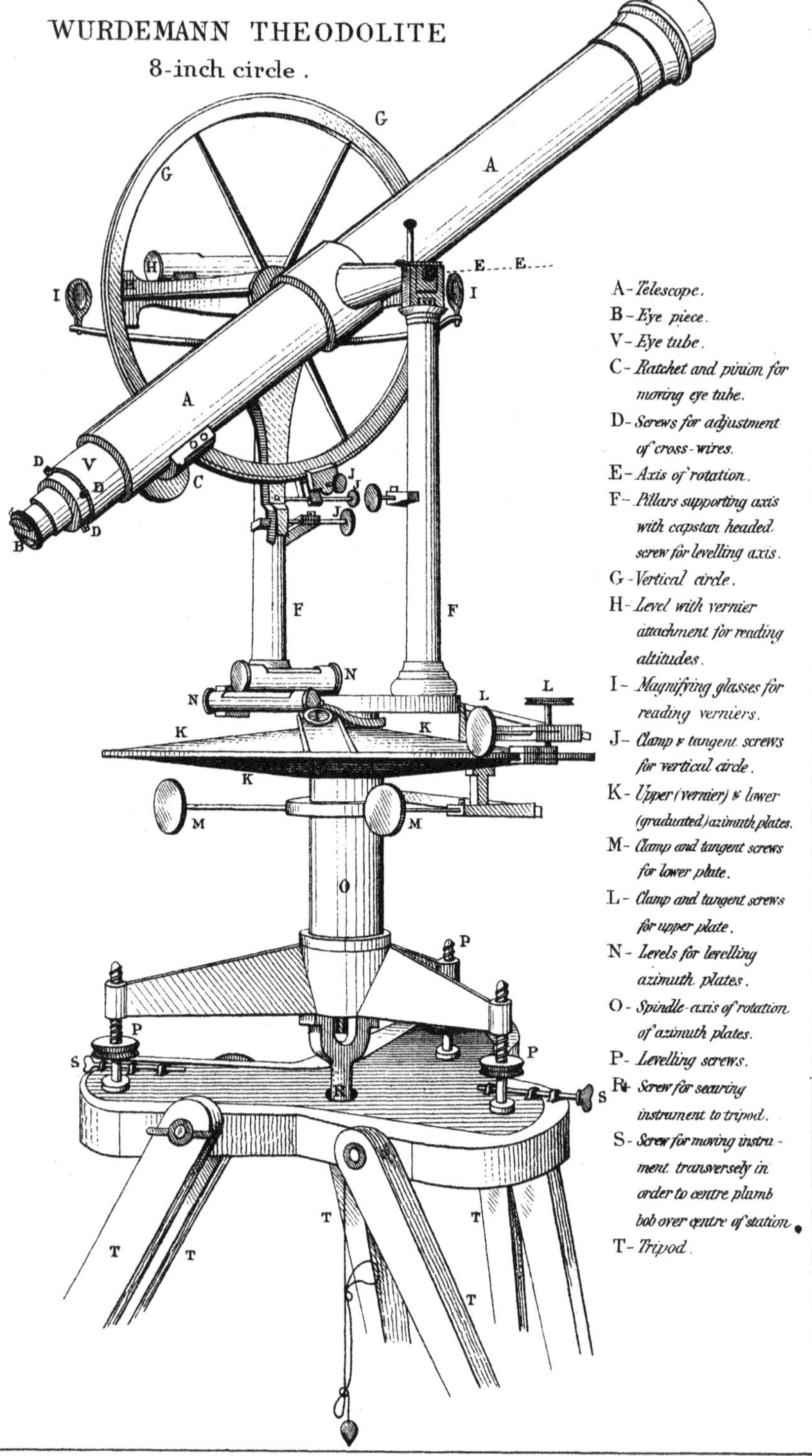

D. Van Nostrand, Publisher, N. Y.

THEODOLITE.

Bureau of Navigation Pattern.

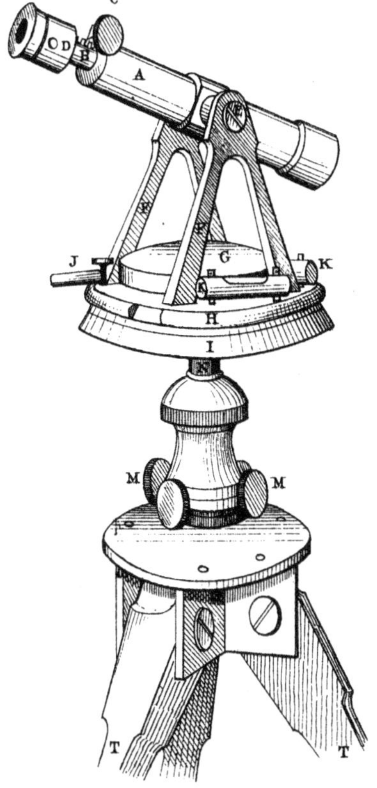

A - *Telescope.*

B - *Eye tube.*

C - *Ratchet & pinion for moving Eye tube.*

D - *Screw for adjustment of cross wires.*

E - *Axis of rotation.*

F - *Pillars supporting axis.*

G - *Compass.*

H - *Upper plate carrying Vernier.*

I - *Lower (graduated) plate.*

J - *Clamp & tangent screws for upper plate.*

K - *Levels.*

M - *Ball & socket joint with 4 levelling screws.*

N - *Spindle - axis of rotation of azimuth plate.*

T - *Tripod.*

D. Van Nostrand, Publisher, N. Y.

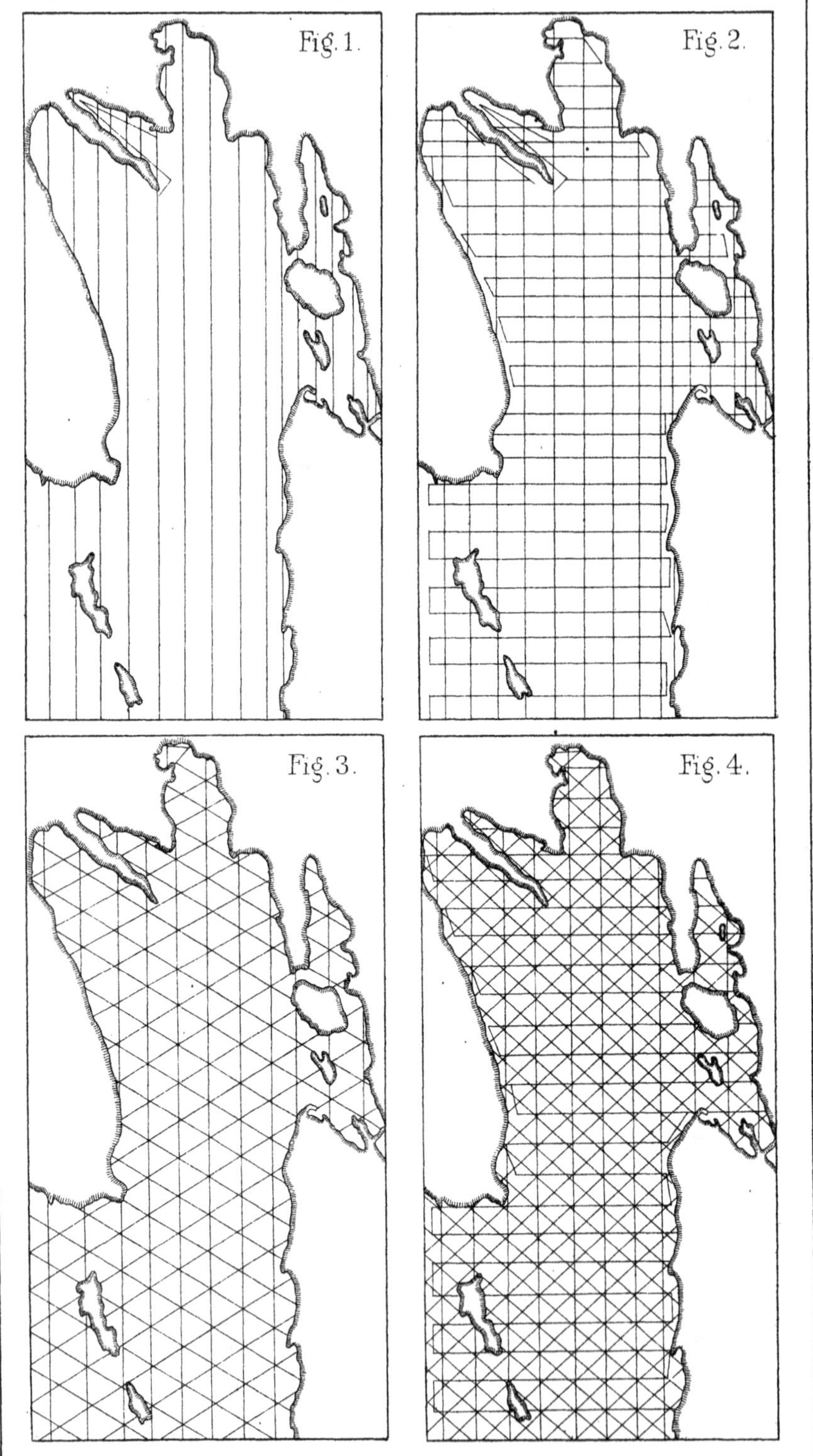

D. Van Nostrand, Publisher, N. Y.

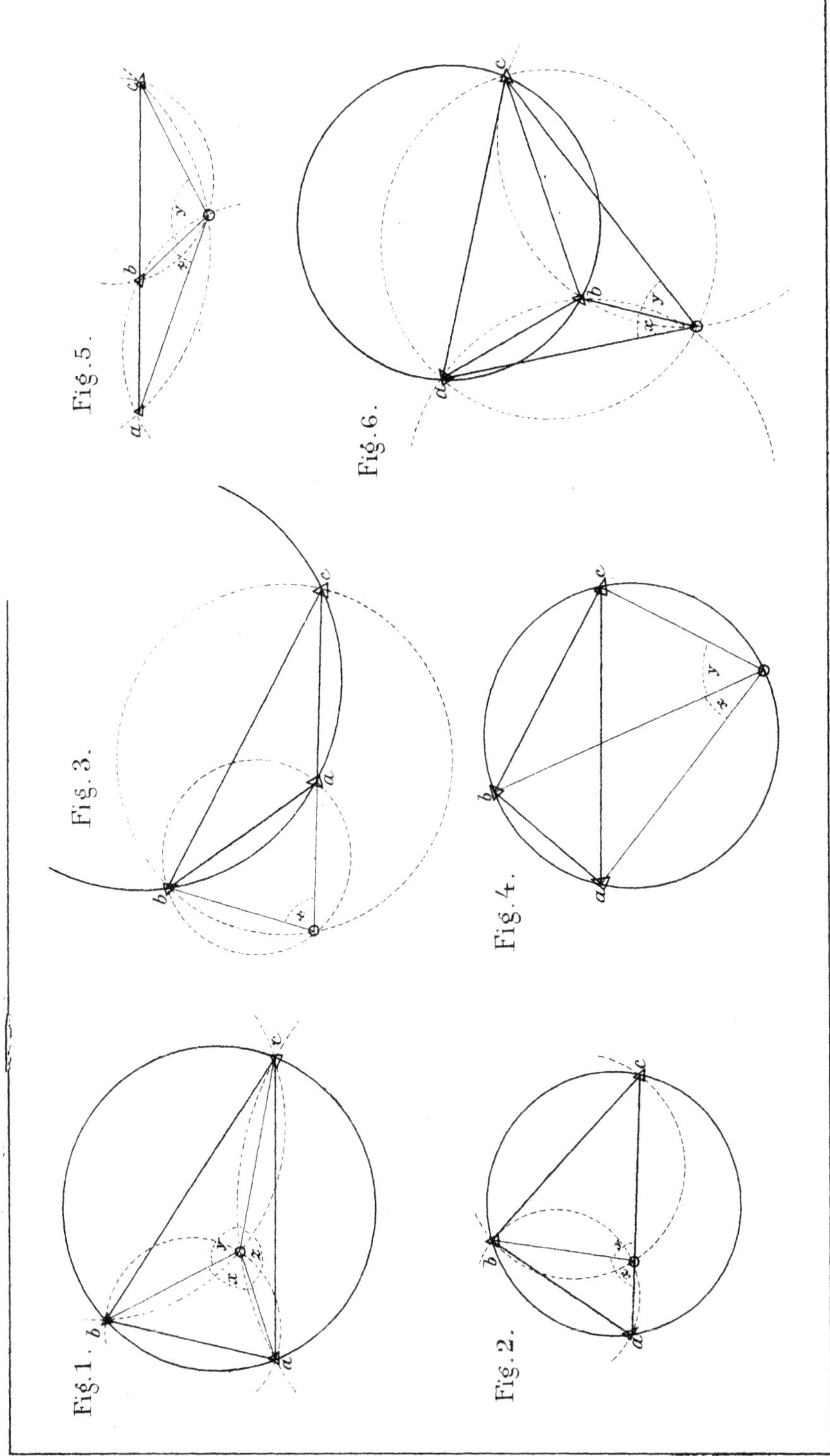

D. Van Nostrand, Publisher, N. Y.

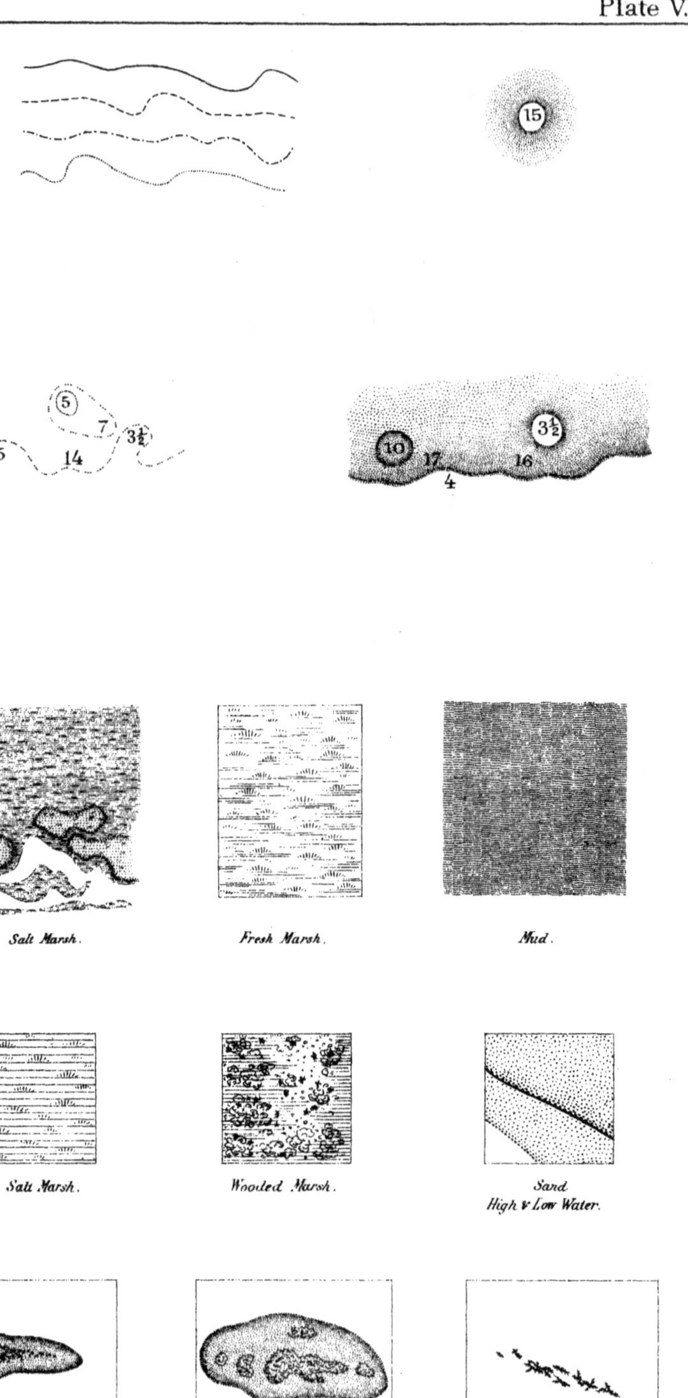

D. Van Nostrand, Publisher, N.Y.

Plate VI.

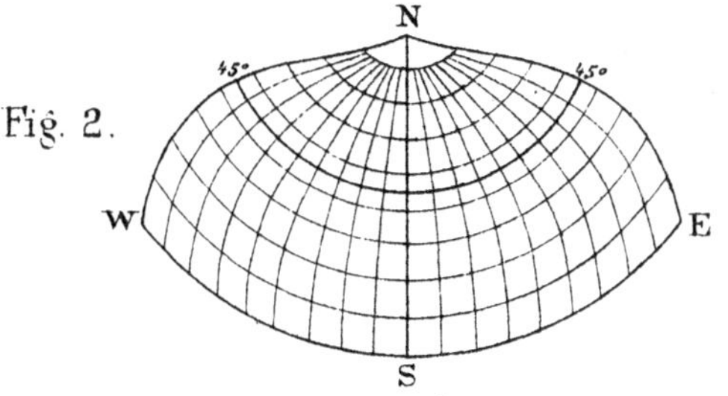

Fig. 2.

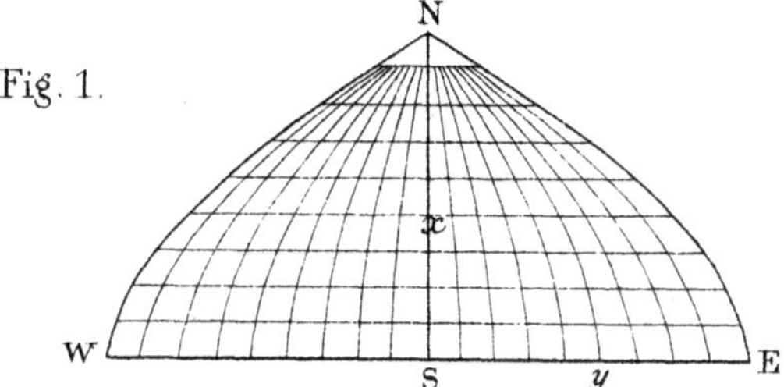

Fig. 1.

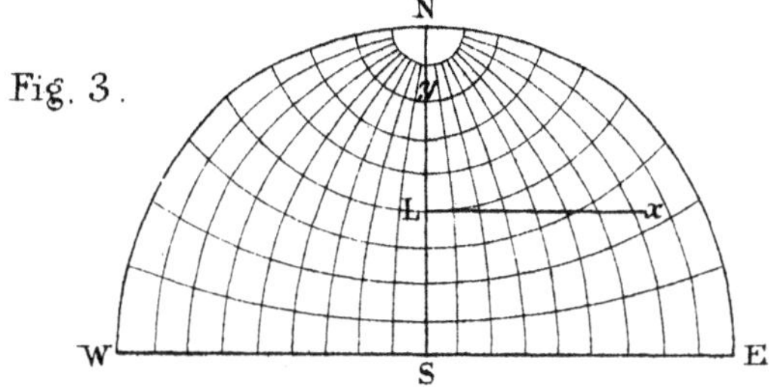

Fig. 3.

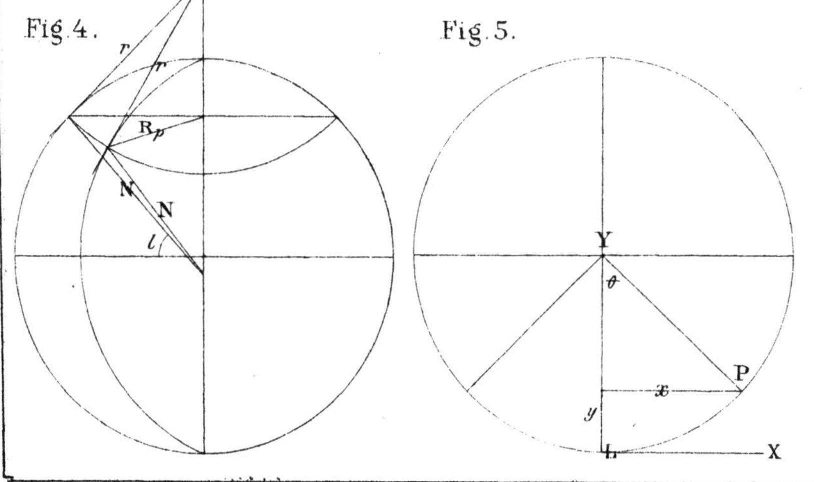

Fig. 4.

Fig. 5.

D. Van Nostrand, Publisher N. Y.

Plate VII.

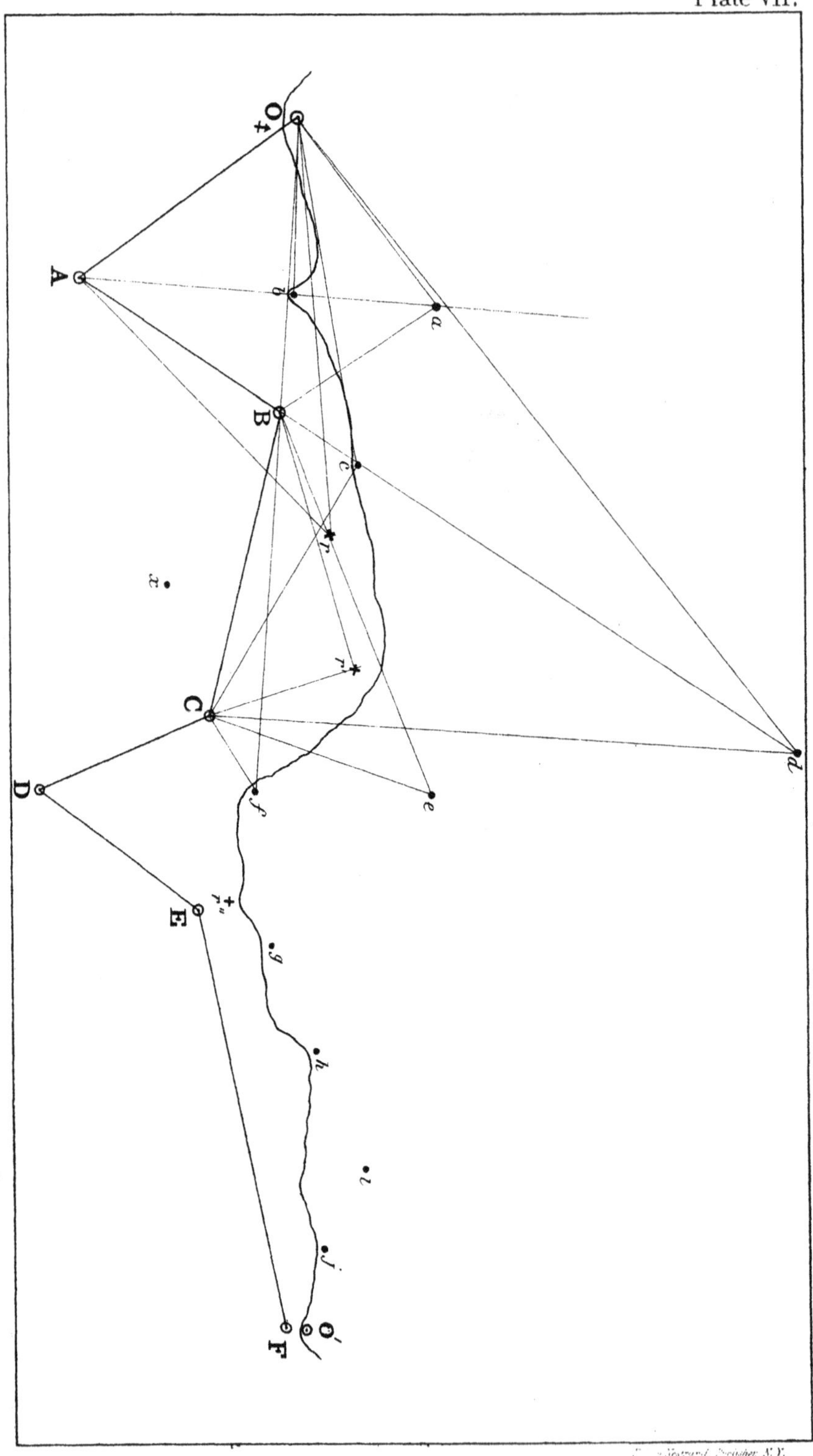

Fig. 1. PORTABLE TRANSIT.

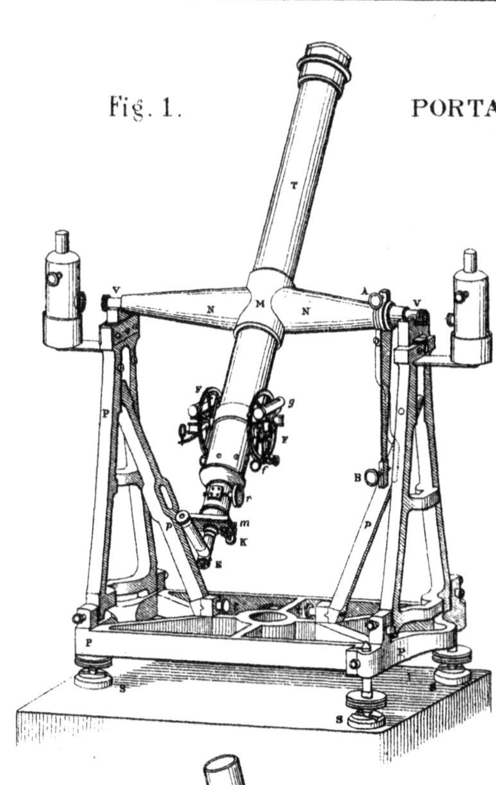

Fig. 2. ZENITH TELESCOPE.

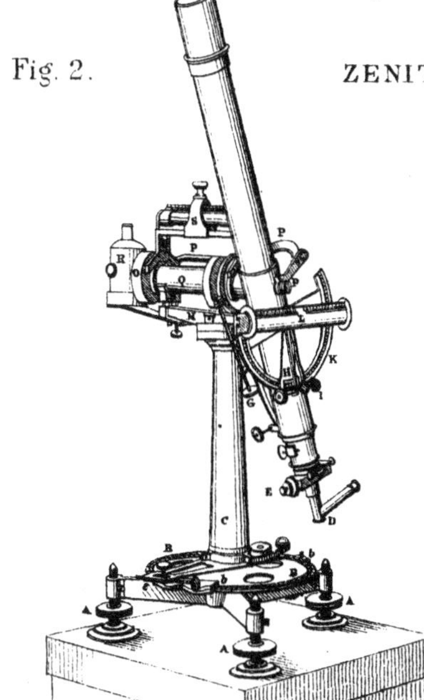

D. Van Nostrand, Publisher N.Y.

Jeffers' Nautical Surveying.

Illustrated with 9 Copperplates and 31 Wood-cut Illustrations. 8vo. Cloth. $5.00.

NAUTICAL SURVEYING. By WILLIAM N. JEFFERS, Captain U. S. Navy.

Many books have been written on each of the subjects treated of in the sixteen chapters of this work; and, to obtain a complete knowledge of geodetic surveying requires a profound study of the whole range of mathematical and physical sciences; but a year of preparation should render any intelligent officer competent to conduct a nautical survey.

CONTENTS.—Chapter I. Formulæ and Constants Useful in Surveying II. Distinctive Character of Surveys. III. Hydrographic Surveying under Sail; or, Running Survey. IV. Hydrographic Surveying of Boats; or, Harbor Survey. V. Tides—Definition of Tidal Phenomena—Tidal Observations. VI. Measurement of Bases—Appropriate and Direct. VII. Measurement of the Angles of Triangles—Azimuths—Astronomical Bearings. VIII. Corrections to be Applied to the Observed Angles. IX. Levelling—Difference of Level. X. Computation of the Sides of the Triangulation—The Three-point Problem. XI. Determination of the Geodetic Latitudes, Longitudes, and Azimuths, of Points of a Triangulation. XII. Summary of Subjects treated of in preceding Chapters—Examples of Computation by various Formulæ. XIII. Projection of Charts and Plans. XIV. Astronomical Determination of Latitude and Longitude. XV. Magnetic Observations. XVI. Deep Sea Soundings. XVII. Tables for Ascertaining Distances at Sea, and a full Index.

List of Plates.

Plate I. Diagram Illustrative of the Triangulation. II. Specimen Page of Field Book. III. Running Survey of a Coast. IV. Example of a Running Survey from Belcher. V. Flying Survey of an Island. VI. Survey of a Shoal. VII. Boat Survey of a River. VIII. Three-Point Problem. IX. Triangulation.

Coffin's Navigation.

Fifth Edition.

12mo. Cloth. $3.50.

NAVIGATION AND NAUTICAL ASTRONOMY. Prepared for the use of the U. S. Naval Academy. By J. H. C. COFFIN, Prof. of Astronomy, Navigation and Surveying, with 52 wood-cut illustrations.

Clark's Navigation.

Illustrated. 8vo. Cloth. $3.00.

THEORETICAL NAVIGATION AND NAUTICAL ASTRONOMY. By Lewis Clark, Lieut.-Commander, U. S. Navy.

Simpson's Ordnance and Naval Gunnery.

Fifth Edition, Revised and Enlarged.

Illustrated with 185 Engravings. 8vo. Cloth. $5.00.

A TREATISE ON ORDNANCE AND NAVAL GUNNERY. Compiled and arranged as a Text-Book for the U. S. Naval Academy, by Commander Edward Simpson, U. S. N.

Young Seaman's Manual.

8vo. Half Roan. $3.00.

THE YOUNG SEAMAN'S MANUAL. Compiled from various authorities, and illustrated with numerous original and select Designs. For the use of the U. S. Training Ships and the Marine Schools.

Harwood's Naval Courts-Martial.

8vo. Law-sheep. $4.00.

LAW AND PRACTICE OF UNITED STATES NAVAL COURTS-MARTIAL. By A. A. Harwood, U. S. N. Adopted as a Text-Book at the U. S. Naval Academy.

Parker's Squadron Tactics.

Illustrated by 77 Plates. 8vo. Cloth. $5.00.

SQUADRON TACTICS UNDER STEAM. By Foxhall A. Parker, Captain U. S. Navy. Published by authority of the Navy Department.

Parker's Fleet Tactics.

18mo. Cloth. $2.50.

FLEET TACTICS UNDER STEAM. By FOXHALL A. PARKER, Captain U. S. Navy. Illustrated by 140 wood-cuts.

Parker's Naval Howitzer Ashore.

26 Plates. 8vo. Cloth. $4.00.

THE NAVAL HOWITZER ASHORE. By FOXHALL A. PARKER, Captain U. S. Navy. With plates. Approved by the Navy Department.

Parker's Naval Howitzer Afloat.

32 Plates. 8vo. Cloth. $4.00.

THE NAVAL HOWITZER AFLOAT. By FOXHALL A. PARKER, Captain U. S. Navy. With plates. Approved by the Navy Department.

Brandt's Gunnery Catechism.

Revised Edition. Illustrated.

18mo. Cloth. $1.50.

GUNNERY CATECHISM. As applied to the service of Naval Ordnance. Adapted to the latest Official Regulations, and approved by the Bureau of Ordnance, Navy Department. By J. D. BRANDT, formerly of the U. S. Navy.

"BUREAU OF ORDNANCE, NAVY DEPARTMENT,
WASHINGTON CITY, July 30, 1864.

"MR. J. D. BRANDT,—

"SIR,—Your 'CATECHISM OF GUNNERY, as applied to the service of Naval Ordnance,' having been submitted to the examination of ordnance officers, and favorably recommended by them, is approved by this Bureau.

I am, Sir, Your obedient servant,

"H. A. WISE, *Chief of Bureau.*"

www.ingramcontent.com/pod-product-compliance
Lightning Source LLC
LaVergne TN
LVHW021405110826
845150LV00007B/1790

* 9 7 8 1 4 2 5 5 1 3 2 8 3 *